Springer Theses

Recognizing Outstanding Ph.D. Research

For further volumes:
http://www.springer.com/series/8790

Aims and Scope

The series "Springer Theses" brings together a selection of the very best Ph.D. theses from around the world and across the physical sciences. Nominated and endorsed by two recognized specialists, each published volume has been selected for its scientific excellence and the high impact of its contents for the pertinent field of research. For greater accessibility to non-specialists, the published versions include an extended introduction, as well as a foreword by the student's supervisor explaining the special relevance of the work for the field. As a whole, the series will provide a valuable resource both for newcomers to the research fields described, and for other scientists seeking detailed background information on special questions. Finally, it provides an accredited documentation of the valuable contributions made by today's younger generation of scientists.

Theses are accepted into the series by invited nomination only and must fulfill all of the following criteria

- They must be written in good English.
- The topic should fall within the confines of Chemistry, Physics, Earth Sciences, Engineering and related interdisciplinary fields such as Materials, Nanoscience, Chemical Engineering, Complex Systems and Biophysics.
- The work reported in the thesis must represent a significant scientific advance.
- If the thesis includes previously published material, permission to reproduce this must be gained from the respective copyright holder.
- They must have been examined and passed during the 12 months prior to nomination.
- Each thesis should include a foreword by the supervisor outlining the significance of its content.
- The theses should have a clearly defined structure including an introduction accessible to scientists not expert in that particular field.

Zefeng Ren

State-to-State Dynamical Research in the F+H$_2$ Reaction System

Doctoral Thesis accepted by
Dalian Institute of Chemical Physics,
Chinese Academy of Sciences, China

 Springer

Author
Dr. Zefeng Ren
State key Laboratory of Molecular
    Reaction Dynamics
Dalian Institute of Chemical Physics,
    Chinese Academy of Sciences
Dalian
People's Republic of China

Supervisor
Prof. Xueming Yang
State key Laboratory of Molecular
    Reaction Dynamics
Dalian Institute of Chemical Physics,
    Chinese Academy of Sciences
Dalian
People's Republic of China

Present address:

International Center for Quantum Materials
Peking University
Beijing
People's Republic of China

ISSN 2190-5053   ISSN 2190-5061 (electronic)
ISBN 978-3-642-39755-4   ISBN 978-3-642-39756-1 (eBook)
DOI 10.1007/978-3-642-39756-1
Springer Heidelberg New York Dordrecht London

Library of Congress Control Number: 2013946339

Printed on acid-free paper

Springer is part of Springer Science+Business Media (www.springer.com)

Parts of this thesis have been published in the following journal articles:

1. Ren, Z. F.; Che, L.; Qiu, M. H.; Wang, X. A.; Dong, W. R.; Dai, D. X.; Wang, X. Y.; Yang, X. M.; Sun, Z. G.; Fu, B. N.; Lee, S. -Y.; Xu, X.; Zhang, D. H., Probing the resonance potential in the F atom reaction with hydrogen deuteride with spectroscopic accuracy. *Proc. Natl. Acad. Sci. USA.* 2008, 105, (35), 12662–12666. (cover article)

2. Che, L.*; Ren, Z. F.*; Wang, X. A.; Dong, W. R.; Dai, D. X.; Wang, X. Y.; Zhang, D. H.; Yang, X. M.; Sheng, L. S.; Li, G. L.; Werner, H. J.; Lique, F.; Alexander, M. H., Breakdown of the Born–Oppenheimer approximation in the F + o-D$_2$ → DF + D reaction. *Science* 2007, 317, (5841), 1061–1064. (*Equal authorship)

3. Ren, Z. F.; Che, L.; Qiu, M. H.; Wang, X. A.; Dai, D. X.; Harich, S. A.; Wang, X. Y.; Yang, X. M.; Xu, C. X.; Xie, D. Q.; Sun, Z. G.; Zhang, D. H., Probing Feshbach resonances in F + H$_2$(j=1) → HF + H: Dynamical effect of single quantum H-2-rotation. *J. Chem. Phys.* 2006, 125, (15), 151102. (JCP Communication)

4. Ren, Z. F.; Qiu, M. H.; Che, L.; Dai, D. X.; Wang, X. Y.; Yang, X. M., A double-stage pulsed discharge fluorine atom beam source. *Rev. Sci. Instrum.* 2006, 77, (1), 3.

5. Qiu, M. H.*; Ren, Z. F.*; Che, L.; Dai, D. X.; Harich, S. A.; Wang, X. Y.; Yang, X. M.; Xu, C. X.; Xie, D. Q.; Gustafsson, M.; Skodje, R. T.; Sun, Z. G.; Zhang, D. H., Observation of Feshbach resonances in the F + H$_2$ → HF + H reaction. *Science* 2006, 311, (5766), 1440–1443. (*Equal authorship)

6. Qiu, M. H.; Che, L.; Ren, Z. F.; Dai, D. X.; Wang, X. Y.; Yang, X. M., High resolution time-of-flight spectrometer for crossed molecular beam study of elementary chemical reactions. *Rev. Sci. Instrum.* 2005, 76, (8), 083107.

Supervisor's Foreword

This thesis describes the scientific achievements of Dr. Zefeng Ren, which were made during his Doctoral Program (2004–2009) in the State Key Laboratory of Molecular Reaction Dynamics, Dalian Institute of Chemical Physics, Chinese Academy of Sciences. During 5 years, a relatively short time as a Researcher, he obtained a number of important results on the State-to-State Reaction Dynamics for the benchmark $F + H_2$ Reaction.

As the Supervisor of his Doctoral Courses, I can introduce two most important findings of his studies. One is the first clear experimental detection of reaction resonances in the $F(^2P_{3/2}) + H_2 \rightarrow HF + H$ reaction, and the other is the direct observation of the breakdown of the Born–Oppenheimer approximation at low collision energies in the $F(^2P_{3/2})/F^*(^2P_{1/2}) + D_2(j = 0) \rightarrow DF + D$ reaction.

Using the H-atom Rydberg tagging time-of-flight crossed molecular beam technique, Dr. Ren studied the reaction resonance and non-adiabatic effects at a full quantum resolved level in the $F + H_2$ system. Through state-to-state resolved experiments, he provided the first conclusive evidence of reaction resonances in the $F(^2P_{3/2}) + H_2 \rightarrow HF + H$ reaction. The dramatic difference between the dynamics for the $F(^2P_{3/2}) + H_2(j = 0, 1)$ reactions represents a textbook example of the role of reactant rotational level in resonance phenomena in this benchmark system. Dr. Ren has also carried out a very high-resolution experimental study on the dynamics of the isotope substituted reaction, $F(^2P_{3/2}) + HD \rightarrow HF + D$, with spectroscopic accuracy (a few cm^{-1}). These findings provided a very clear physical picture for reaction resonances in this benchmark system, which has eluded us for more than 30 years.

In this thesis work, Dr. Ren also studied the non-adiabatic effect in the $F + D_2$ reaction, where the $F^*(^2P_{1/2})$ is expected to be non-reactive according to the Born–Oppenheimer approximation. He measured accurately the population ratio of $F(^2P_{3/2})$ and $F^*(^2P_{1/2})$ in the beam using synchrotron radiation single photon autoionization, then determined the relative reactivity of F and F* with D_2. For the first time, he found that $F^*(^2P_{1/2})$ is more reactive than $F(^2P_{3/2})$ at low collision energy, providing a clear case of the breakdown of Born–Oppenheimer approximation. This is the first accurate experimental measurement of the non-adiabatic effects of this important system.

These results, reported in the Journals of *Science*, *PNAS*, and *JCP*, give us deep insights into how various elementary chemical reactions actually occur. The thesis makes an important contribution toward our understanding of the nature of the chemical reaction. The relevance of the issues, the soundness of hypotheses, the data set, and the analysis are key strengths of this work. I recommend this excellent study to researchers on molecular reaction dynamics as an inspiring and enlightening piece of literature.

I hope that there will be many grateful readers who have gained a broad perspective of molecular reaction dynamics as a result of the author's efforts.

Dalian, China, May 2013   Xueming Yang

Acknowledgments

First and foremost, I would like to thank my Supervisor, Prof. Xueming Yang. With his active academic thinking and energy, he has not only accurate insight into scientific frontiers, but also experienced experimental techniques. I sincerely appreciate his invaluable academic and personal support I received from Prof. Yang throughout the work. During 5 years in Dalian, I got a lot of opportunities and good scientific exercises, which are critical for my scientific carrier. In addition, Prof. Yang's interest and passion for scientific research also deeply affect me.

During the work, Prof. Dongxu Dai gave me valuable guidance, especially during the experiments. Without his support and encouragement, this work would not have been possible. I also respectfully acknowledge Prof. Xiuyan Wang, Zichao Tang, and Senior Engineer Bo Jiang, for the regular guidance and valuable discussions.

I would like to express my heartfelt thanks to Minghui Qiu, Li Che, Xingan Wang, Wenrui Dong, Chuanyao Zhou, and Zhibo Ma, for the help of the completion of my thesis. Gratitude is also extended to my lovely colleagues in group 1102, Qing Guo, Chunlei Xiao, Chengbiao Xu, Kaijun Yuan, Weiqing Zhang, Yongwei Zhang, Yuan Cheng, Zhichao Chen, Huilin Pan, Fengyan Wang, Quan Shuai, Lina Cheng. I thank all of you for creating a comfortable atmosphere, for the technical assistance, for working together, and so many other things which I cannot name all.

I am very thankful to Prof. Donghui Zhang, Prof. Daiqian Xie, and Prof. Alexander for the theoretical calculation, which gives correct interpretation of our experimental results and reveals the dynamics picture deep behind the phenomenon. Thanks to Prof. Kopin Liu from Institute of Atomic and Molecular Science, Academia Sinica, Taiwan, for his helpful discussion. Thanks to Prof. Liusi Sheng, Prof. Xiaobin Shan, and Sisheng Wang from National Synchrotron Radiation Laboratory, University of Science and Technology of China, for their technical assistance in the synchrotron radiation experiments. Thanks to Prof. Heping Yang, Wenbo Shi from group 1105 for the help in measurement of the purity of para-H_2 and ortho-D_2.

Thanks to all teachers and students from group 11, especially Fu'e Li, Yongzheng Song, Wei Wu, and Xiaoying Fan, for their concern and help in daily life and at work.

Finally, I would like to deeply thank my parents, my wife, and my son. Thank you for your love, support, and patience. I am truly blessed to have you as my family.

Contents

Chapter 1
Introduction

The nature of chemical reactions is the breaking of old bonds and the formation of new bonds. Molecular reaction dynamics is the study on how the old bonds are broken and how the new bonds are formed. In the past few decades, molecular reaction dynamics is an important field of physical chemistry and chemical physics. Its main task is to study elementary chemical reaction processes on the atomic scale and femtosecond (even attosecond) time scale. The in-depth study of this field offers important knowledge to atmospheric chemistry, interstellar chemistry, as well as combustion chemistry, and deepens our understanding of the essential nature of chemical reactions in nature. In the 1980s, the improvement of the crossed molecular beam technique [1] and femtosecond chemistry [2] enabled the molecular reaction dynamics to have a rapid development, and gradually mature. In the last ten years, the molecular reaction dynamics had a lot of progress and made a series of achievements, due to the many new technologies, the development, and innovation of new experimental methods, and advances in theory and computation [3].

This thesis focuses on experimental study of the resonance phenomenon and the breakdown of the Born–Oppenheimer (B–O) approximation in the $F + H_2$ and its isotopic reactions. This chapter introduces briefly two basic concepts: (1) the reaction resonance; (2) the B–O approximation.

1.1 Basic Concepts of Reaction Resonance

The resonance effect was first observed on the pendulum by Galileo in 1602. When we play swing, if we push the swing with the swing interval (that is, the so-called resonance frequency), then the swing will be getting higher and higher; if we push slightly faster or slower, the swing will swing only a small arc. This is the resonance effect found 400 years ago. In daily life, many things are based on resonance effect, such as musical instruments, quartz watch, radio and television signal, and so on. Also, many times, we need to avoid the resonance effect, for instance, soldiers are ordered to break their steps while crossing a bridge.

Z. Ren, *State-to-State Dynamical Research in the F+H₂*
Reaction System, Springer Theses, DOI: 10.1007/978-3-642-39756-1_1,
© Springer-Verlag Berlin Heidelberg 2014

Resonance is a very common physical phenomenon, ubiquitous from nuclear physics to chemical reactions [4]. The most familiar example comes from spectroscopy. When the photon energy hν is equal to the energy difference of the two energy levels of the atom or molecule ΔE_{12}, then the atom or molecule will absorb the photon. Absorption spectrum shows a sharp peak, the width of which is associated with the excited-state lifetime. This phenomenon is not just confined to the photon. In the 1930s, Fermi and collaborators found a great cross section in slow neutron scattering at some energy. In the following years, the study of elementary particle resonance phenomenon played an important role in understanding of the fundamental forces, for instance, pion-nucleon elastic scattering at low energy. In the 1950s, experimentalists and theorists almost simultaneously discovered resonance phenomenon in the electron-atom/molecule scattering. For example, in e-N_2 scattering, the fine structures of the collision cross section correspond to the vibrational excitation of N_2 molecules. In the late 1960s, scientists began to study the contribution of resonance to the cross section in the Cs + RbCl reaction. In their scattering experiments, Herschbach et al. found that this reaction experiences a long-lived intermediate state. In the early 1970s, Truhlar and Kuppermann elucidated the importance of the scattering resonance in the chemical reaction more clearly. They calculated the H + H_2 collinear configuration accurately for the first time, using quantum mechanics [5, 6]. They found that the overall reaction and non-reaction cross sections exhibit significant oscillation phenomenon with collision energy. They attributed this characteristic to the interference between different semi-classical channels of the reactants and products. Levine and Wu made close-coupling linear calculation on this system and clearly demonstrated resonance in this reaction. This surprising phenomenon is because of a potential well in the H + H_2 potential energy surface (PES), which can be bound to a long-lived intermediate state, although later it is confirmed that this potential well does not exist. In the past few decades, scientists had wide-ranging and in-depth research on the nature's most simple chemical reaction H + H_2 and found barrier resonance in this reaction [7, 8]. Another benchmark reaction resonance study is the F + H_2 system, which is a typical example and the main work of this thesis.

Eyring and Polanyi raised a very important concept—the transition state in 1935. The transition state is an intermediate state from reactants to products in a chemical reaction. Direct observation of the transition state is considered to be Holy Grail of chemistry. A typical chemical reaction has a simple energetic barrier (Fig. 1.1a). There is no discrete quantum state in the transition-state region along the reaction path. When the reaction energy excesses the energetic barrier, the reaction probability shows a simple increment (Fig. 1.1b). In the time delay curve in Fig. 1.1b, a very broad peak can be observed when the reaction energy is equal to the energetic barrier. Simple to understand, the movement of the reactants in the reaction path is more and more slow with increasingly high-potential energy. However, in some cases, there is a potential well along the minimum energy path, in which there are some bound quantum states, called dynamical resonance states, or reaction resonance states, as shown in Fig. 1.1c. Due to the presence of

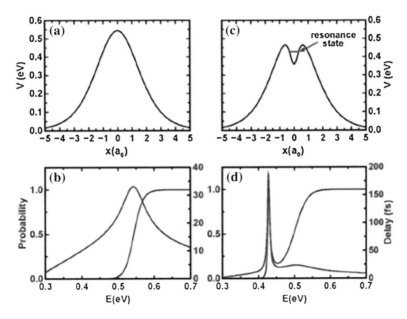

Fig. 1.1 One-dimensional views of two model reactions, a reaction with a simple barrier and a typical reaction with a dynamical resonance: **a** the potential energy curve along the reaction coordinate for a model reaction with a simple barrier; **b** the calculated reaction probability and time delay for the model reaction in panel **a**; **c** the potential energy curve along the reaction coordinate for a model reaction with a dynamical resonance; **d** the calculated reaction probability and time delay for the model reaction in panel **c**. The two reaction models are adapted from ref. [22]. Reprinted with the permission from Ref. [23]. © 2008 American Chemical Society

resonance states, the reaction probability has a very sharp peak, when the reaction energy exactly matches the reaction-resonance-state energy, lower than the maximum PES (barrier), as shown in Fig. 1.1d. At reaction resonance energy, there is a very sharp peak of the time delay curve, which means that the reaction transition states can exist for a long time in this potential well, and the life is usually hundreds of femtoseconds (Fig. 1.1d). When the reaction energy is equal to the potential barrier height, there is a wide protrusion in the time delay curve, caused by the same reason as in Fig. 1.1b.

Reaction resonance is a kind of transition state. PES is very sensitive to the reaction resonance. Studying reaction resonance can accurately test and probe the structure of the potential energy surface, and thus the nature of the chemical reaction is better to understand. However, probing the reaction resonance is not an easy but very challenging work, both experimentally and theoretically. Thus, in the past few decades, the reaction resonance has been a major research direction of molecular reaction dynamics [9].

There are three categories of reaction resonance: shape resonance, Feshbach resonance, and barrier resonance [4, 10].

Fig. 1.2 Schematic of shape
resonance mechanism

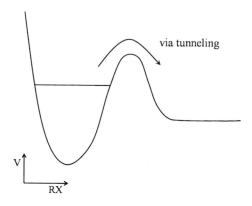

Figure 1.2 is a schematic of shape resonance mechanism [11]. Adding a centrifugal potential on a simple potential energy surface, centrifugal barrier and potential well will be formed. Certain quantum states can exist in a potential well, and the system can be bound in a one-dimensional potential well, which is a shape resonance. The system has a very long lifetime in the potential well and can escape from the potential well by tunneling effect. Shape resonance is one-dimensional, whose reaction coordinate degrees of freedom do not couple with other systems. Shape resonance can impact and control the product scattering angular distribution as well as other dynamical effects.

In Feshbach resonance system, the reaction coordinate couples with the other freedom degrees. The system can form a quasi-bound or metastable complex, which is Feshbach resonance, also known as dynamic resonance. Even in pure repulsion potential energy surface, this resonance state can be formed. Quasi-bound complex can react with other adiabatic potential surface, producing non-adiabatic coupling dissociation product (Fig. 1.3) [11]. In F + H_2 reaction, Feshbach resonance results forward scattering, producing HF ($v' = 2$), see Chap. 3 of this thesis.

Barrier resonance, also known as threshold resonance, Fig. 1.4 is a schematic of barrier resonance of a triatomic system. In transition-state region, there are the other two vibration directions along the vertical direction of the reaction path (three-atom system as an example). In this direction, there are a number of states—quantum bottleneck states, defined by two quantum numbers. In the process of climbing along the PES of the system, the system kinetic energy is converted to the internal energy. When the reaction energy is higher than the potential barrier, products can be produced. The reaction energy becoming larger, the system can achieve different quantum bottleneck states, which open new reaction channels, thereby affecting the reaction probability. Moore et al. found steplike photolysis rate with increasing light energy in Ketene photolysis experiments $C_2H_2O + h\nu \rightarrow CO + CH_2$ [12]. The authors believed that the increasing photolysis light energy opens different quantum bottleneck states. Xueming Yang and Skodje et al. clearly observed dynamic effects of quantum bottleneck states in H + $D_2 \rightarrow$ HD + D crossed molecular beam studies. Theoretical calculation and analysis further clarified the characteristics of quantum bottleneck states. Barrier resonance has different

Fig. 1.3 Schematic of
Feshbach resonance
mechanism

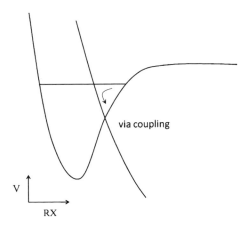

Fig. 1.4 Schematic of
barrier resonance mechanism

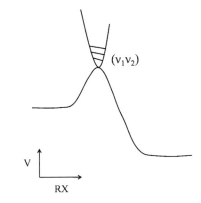

performance from the first two resonances, and the reaction probability does not exhibit the characteristics of resonance peaks, so it is not classified as resonance by some researchers [7].

Sometimes, it is difficult to accurately say resonance effect of a chemical reaction system is from which of the above-mentioned mechanisms. The difference between them is sometimes not very clear. The $F + H_2$ reaction product HF ($v' = 3$) forward scattering found by Yuan T. Lee et al. [13] was presumably from the Feshbach resonance. Experimental and theoretical research in recent years attributed the forward scattering to time delay effect and shape resonance [14].

1.2 Basic Concepts of Born–Oppenheimer Approximation

B–O approximation, also known as the adiabatic approximation, plays a very important role in many fields of physics and chemistry. The masses of electrons are much lighter than those of nucleus, thus electrons move much faster than nucleus. With small movements of nucleus, the electrons are able to quickly adjust their states of movement in order to adapt to the new nuclear potential field. Oppositely, the nucleus is not sensitive to the changes of electrons. Based on this fact, Born and Oppenheimer raised the concept of B–O approximation in 1927. Within B–O approximation, the movements of the nucleus and the electrons are separated. When dealing with the movements of electrons, the nucleus is considered to be still; when dealing with the movements of nuclear, the electron energy is considered as potential energy of the nuclear movements. This is also the basic idea to establish a potential energy surface. Within B–O approximation, the nucleus moves only on an adiabatic potential energy surface. In other words, the electron movement and nuclear movement are decoupled. In 1935, Michael Polanyi and Henry Eyring realized the B–O approximation can greatly simplify multibody Schrödinger equation and established a PES using the B–O approximation, to study the $H + H_2 \rightarrow H_2 + H$ reaction.

Although the B–O approximation has a very wide range of successful applications, but lots of physical and chemical processes cannot be well described with B–O approximation. Because when two or more electronic states degenerate or come close, B–O approximation will break down, and there will be a non-adiabatic process, like molecular predissociation [15], charge transfer [16], chemical adsorption of gas molecules on surfaces [17], scattering of high-vibrational excited molecules on surface[18], and many chemical processes on the metal surface [19]. Although the B–O approximation breaks down in a lot of systems, and non-adiabatic processes even play a decisive role in complex chemical reactions, it is not easy to determine how large this non-adiabatic coupling is, experimentally or theoretically. So, the non-adiabatic coupling is a very challenging topic in the research of molecular reaction dynamics in recent years and also a research focus [20, 21]. In Chap. 4 of this thesis, the non-adiabatic effects of the $F + D_2 \rightarrow DF + D$ reaction are described in detail, and the strength of the non-adiabatic coupling is determined.

This thesis will be organized as follows: The Chap. 2 describes the H atom Rydberg tagging time-of-flight crossed molecular beam apparatus and experimental methods in our laboratory; the Chap. 3 shows the research of resonance phenomenon in the $F + H_2$ reaction; in the Chap. 4, the breakdown of the B–O approximation in the $F + H_2$ reaction is introduced.

References

1. Lee YT (1986) Molecular beam studies of elementary chemical processes. In: Nobel Lecture
2. Zewail AH (1999) Femtochemistry: atomic-scale dynamics of the chemical bond using ultrafast lasers. In: Nobel lecture
3. Yang X, Liu K (eds) (2004) Modern trends in chemical reaction dynamics: Experiment and theory. World Scientific Singapore, Singapore
4. Fernandez-Alonso F, Zare RN (2002) Scattering resonances in the simplest chemical reaction. Annu Rev Phys Chem 53:67–99

5. Truhlar DG, Kuppermann A (1972) Exact and approximate quantum mechanical reaction probabilities and rate constants for the collinear $H + H_2$ reaction. J Chem Phys 56:2232–2252
6. Truhlar DG, Kuppermann A (1970) Quantum mechanics of the $H + H_2$ reaction: exact scattering probabilities for collinear collisions. J Chem Phys 52:3841–3843
7. Dai DX, Wang CC, Harich SA et al (2003) Interference of quantized transition-state pathways in the $H + D_2 \rightarrow D + HD$ chemical reaction. Science 300:1730–1734
8. Harich SA, Dai DX, Wang CC et al (2002) Forward scattering due to slow-down of the intermediate in the $H + HD \rightarrow D + H_2$ reaction. Nature 419:281–284
9. Friedman RS, Truhlar DG (1991) Chemical reaction thresholds are resonances Chem Phys Lett 183:539–546
10. Schatz GC (2000) Reaction dynamics: detecting resonances. Science 288:1599–1600
11. Liu KP (2001) Crossed-beam studies of neutral reactions: state-specific differential cross sections. Annu Rev Phys Chem 52:139–164
12. Child MS (1974) Molecular collision theory. Academic Press, London & New York
13. Kim SK, Lovejoy ER, Moore CB (1995) Transition-state vibrational level thresholds for the dissociation of triplet ketene. J Chem Phys 102:3202–3219
14. Neumark DM, Wodtke AM, Robinson GN et al (1985) Molecular beam studies of the $F + H_2$ reaction. J Chem Phys 82:3045–3066
15. Wang XG, Dong WR, Qiu MH et al (2008) HF (v '= 3) forward scattering in the $F + H_2$ reaction: shape resonance and slow-down mechanism. Proc Natl Acad Sci USA 105:6227–6231
16. Butler LJ (1998) Chemical reaction dynamics beyond the Born-Oppenheimer approximation. Annu Rev Phys Chem 49:125–171
17. Closs GL, Miller JR (1988) Intramolecular long-distance electron transfer in organic molecules. Science 240:440–447
18. Gergen B, Nienhaus H, Weinberg WH et al (2001) Chemically induced electronic excitations at metal surfaces. Science 294:2521–2523
19. White JD, Chen J, Matsiev D et al (2005) Conversion of large-amplitude vibration to electron excitation at a metal surface. Nature 433:503–505
20. Frischkorn C, Wolf M (2006) Femtochemistry at metal surfaces: nonadiabatic reaction dynamics. Chem Rev 106:4207–4233
21. Wodtke AM (2006) Chemistry in a computer: advancing the in silico dream. Science 312:64–65
22. Polanyi JC, Zewail AH (1995) Direct observation of the transition-state. Acc Chem Res 28:119–132
23. Yang XM, Zhang DH (2008) Dynamical resonances in the fluorine atom reaction with the hydrogen molecule. Acc Chem Res 41:981–989

Chapter 2
Hydrogen Atom Rydberg Tagging Time-of-Flight Crossed Molecular Beam Apparatus

Hydrogen is the most abundant and most important element in the universe. The hydrogen is the simplest atom and also the lightest and composed of a proton and an electron. Therefore, hydrogen atom is intensively studied. In nature, hydrogen-containing compounds are very important. Photochemical and chemical reactions of the hydrogen-containing compound are important subjects in the molecular reaction dynamics. For instance, photodissociation of water $H_2O + h\nu \rightarrow OH + H$ [8, 39], photodissociation of formaldehyde $CH_2O \rightarrow CHO + H$ [34, 38], the simplest bimolecular chemical reaction in nature $H + H_2 \rightarrow H_2 + H$ [1, 7, 11], and the famous reaction $F + H_2 \rightarrow HF + H$ [18]. This chapter focuses on hydrogen atom Rydberg tagging time-of-flight (HRTOF) crossed molecular beam apparatus, which was used for the research described in this thesis. In Sect. 2.1, I will introduce the general knowledge of molecular beam and crossed molecular beam and some history; HRTOF technique is described in Sect. 2.2; the vacuum system, detection and acquirement system as well as the resolution of the apparatus are described in detail in Sect. 2.3.

2.1 Molecular Beam and Crossed Molecular Beam Techniques

2.1.1 Molecular Beam Technique

Duniyer carried out the first molecular beam experiment in 1911. By that time, high-speed vacuum pump had just been invented, so a good vacuum could be achieved to avoid collisions between molecular beam and background gases. Otto Stern was the first scientist who carried systematic research on molecular beams. He measured the Maxwell–Boltzmann distribution of the atom speed in the silver atom beam. Isidor Rabi created a new era of molecular beam, by carrying out magnetic resonance studies using molecular beam techniques [12]. Figure 2.1 shows the history of the development of molecular beam technique.

Z. Ren, *State-to-State Dynamical Research in the F+H2*
Reaction System, Springer Theses, DOI: 10.1007/978-3-642-39756-1_2,
© Springer-Verlag Berlin Heidelberg 2014

Fig. 2.1 Development
history of molecular beam
technique and other close
techniques, cited from the
Nobel Lecture of Herschbach
in 1986 [12]

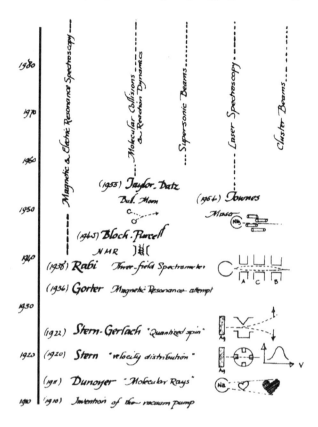

The earliest molecular beam source is a small hole at one side of a container containing reactants, where molecules flow effusively. The molecular mean free path of the beam source (λ_o) is much larger than the size (D) of the furnace orifice, so that the molecules outflow without collision. This is the so-called effusive source or diffusion source. This is the method Stern used to measure the silver atom speed at the very beginning. There is no collision between molecules, so the molecular speed distribution is the Boltzmann distribution at the corresponding reservoir temperature. Gas molecular flow will be generated if condensable reactants pass through heating container. The advantage of effusive source is that it is applicable to a lot of material, and its structure is simple and easy to be controlled. However, it is required that the molecular mean free path is much larger than the orifice size, which limits the gas pressure of the beam source chamber; thus, a molecular beam of high intensity cannot be formed, and the speed distribution is also very wide.

In 1951, Kantrowitz and Grey recommended using a supersonic jet as a molecular beam source [13, 14]. The beam source and orifice design were changed to D $\gg$ λ_o; thus, the gas molecules have a strong collision at the orifice and downstream. During the expansion process, heat energy of the random movement is converged into kinetic energy of molecular directional movement; this process is to accelerate the molecular velocity (u) in the molecular beam. The random movement becomes orderly directional movement, so that the temperature of the molecular beam is reduced, and the local sound velocity ($a = (\gamma kT/m)^{1/2}$, in which γ is the heat capacity ratio C_p/C_v, k is Boltzmann constant) is also declined, which is proportional to $T^{1/2}$, and the Mach number ($M = u/a$) is continuously increased. This is why this type of molecular beam is called supersonic molecular beam [32]. This supersonic molecular beam technique not only makes the molecular beam strength increased by several orders of magnitude, to reduce the temperature of the molecular beam, but also narrows molecular speed distribution (reduced molecular speed only means translational temperature is low).

Figure 2.2 is the velocity distribution in effusive molecular beam and supersonic molecular beam for helium. The molecular velocity from effusive source follows the Boltzmann distribution, which is very wide. The molecular velocity distribution from a supersonic molecular beam is much narrower, as shown in Fig. 2.2, assuming the speed ratio $v/\Delta v = 10$. In recent years, Even et al. obtained molecular beams with the speed ratio greater than 100, with improvement in the pulse valve. The temperature of the molecular beams reaches a temperature of mK; thus, He clusters can be produced [9]. Thomas et al. obtained molecular beams with speed ratio greater than 1,000 [35]. Due to the constant collisions between the molecules in the adiabatic expansion, a portion of the molecular rotational and vibrational energy can also be converted into translational kinetic energy, reducing the molecular rotational and vibrational temperature. Rotation–translation relaxation efficiency is relatively high, so it is possible to get a relatively low temperature of rotational distribution, which makes state-to-state kinetics (i.e., products of certain quantum states generated by reactants of certain quantum states) study possible. Vibration–translation relaxation efficiency and vibration–rotation relaxation efficiency are low, so the vibrational temperature is difficult to be cooled. Scientists

Fig. 2.2 Velocity distribution from effusive molecular beam (*dot*) and supersonic molecular beam (*solid line*). Both curves are normalized to unity at the most probable velocity and are for helium at a reservoir temperature of 300 K. The curve for the supersonic molecular beam assumes the speed ratio $v/\Delta v = 10$

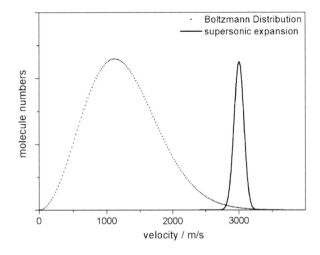

take advantage of the characteristics of vibrational relaxation to study the effect of some vibrational excitation on the reactions [36].

Molecular beam speed can be easily changed by the incorporation of a different carrier gas. For instance, when the heavier molecules are mixed with a large number of lighter carrier gas, such as He or H_2, the heavier molecules can be accelerated; when the lighter molecules are mixed with a large number of heavier carrier gas such as Ar, the lighter molecules can be decelerated.

The supersonic molecular beam technique plays an invaluable role in the development of modern physical chemistry and has contributed greatly to the molecular reaction dynamics [30, 31]. Nowadays, molecular beam technique is indispensable in the fields of photolysis, crossed molecular beam, cluster and molecular beam–surface scattering.

2.1.2 Crossed Molecular Beam Technique

Scattering method plays a very important role in the development of modern physical chemistry. In 1909, Ernest Rutherford put forward a model of atomic structure by observing the α-particle scattering experiments on the gold foil. Scattering becomes a good tool to probe structure of the microscopic world and dynamics. The main basis for judging the collision interaction in the collision process is the angular distribution after collision. In crossed molecular beam experiment, the chemical reaction dynamics is studied by detecting product species, state and angular distribution, after collision scattering of two atom beams or molecular beams. So far, crossed molecular beam is the most powerful tool and method in the study of molecular reaction collision.

The first crossed molecular beam experiment was carried out in 1955 at the Oak Ridge National Laboratory by Taylor and Datz [33]. Later, Herschbach and Yuan T. Lee improved the crossed molecular beam apparatus and carried out a series of ground-breaking study of the crossed molecular beam experiments. Because of this, they won the 1986 Nobel Prize in Chemistry. Early crossed molecular beam experiments were mostly used to study alkali metal reactions, limited by the detection method, surface ionization. When the scattering alkali metal atoms hit the hot filament, they are ionized to produce a small current, which can be detected. This detection method is simple, but the efficiency is very high and very sensitive. However, it can only be used to detect an alkali metal and cannot detect other molecules or atoms. Therefore, this era is also known as alkali age [12]. Surface ionization can give good information on angular distribution, but no information on translational energy. In order to obtain product translational energy distribution, early crossed beam apparatus used two tunable with a slit between molecular beam crossover center and detectors. By controlling the rotational speed of the two slits, each time only product with a certain speed can pass through and be detected. Between 1967 and 1968, Yuan T. Lee and co-workers obtained product translational energy by time-of-flight mass spectrometry, using electron impact ionization and quadruple mass spectrometry to choose mass/charge.

Electron impact ionization can be applied to almost all neutral molecules and atoms. In addition, the background gas pressure in ionization region can be reduced using differential pumping, to improve the signal-to-noise ratio of the detection, so such apparatus is known as universal [18, 19]. At that time, this apparatus was named "Hope" in their laboratory [12]. In the following decades, the universal crossed molecular beam apparatus not only brings "Hope" to molecular reaction dynamics, but also has become a powerful weapon for human to explore the nature of chemical reactions. Yang et al. improved vacuum degree in electron bombardment region using cryopump, inhibiting the H_2 background and making measurement of H_2 or H product channels possible [20]. Yang et al. also used tunable VUV light generated by synchrotron radiation as the ionization source, avoiding background of some product channels caused by broken product molecules in electron impact ionization [40]. Casavechia achieved a so-called "soft ionization" by controlling the energy of electron impact ionization [6].

With good supersonic molecular beam, it is no longer difficult to achieve crossed molecular beam. The difficulty is how to effectively detect the translational and ro-vibrational states of the product, as well as the angular distribution of the product. Besides the above-mentioned methods, scientists developed a lot of new technology, which greatly promoted the development of molecular reaction dynamics.

2.1.3 Laboratory Coordinate System and the Center-of-Mass Coordinate System Conversion

In crossed molecular beam experiments, measurements of the product angle and speed are taken in the laboratory coordinate system (LAB). But information in the center-of-mass coordinate system is required to explain the dynamics of the scattering process. Thus, the results obtained in the laboratory coordinate system must be transformed to the center-of-mass coordinate system. We usually use the Newton vector diagram, i.e., the velocity vector diagram.

Figure 2.3 is the Newton diagram for the reaction $A + BC \rightarrow AB + C$. An atom beam with a speed V_A and the BC molecule (or radical) beam with a speed V_{BC} cross and collide, generating products AB and C. Before the reaction, the relative velocity and the center-of-mass velocity of the A and BC are, respectively,

$$V = V_A - V_{BC} \tag{2.1}$$

$$V_{C.M.} = \frac{m_A V_A + m_{BC} V_{BC}}{m_A + m_{BC}} \tag{2.2}$$

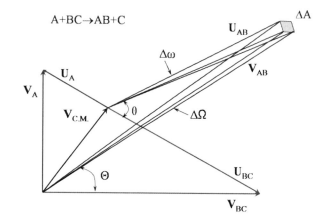

Fig. 2.3 Newton vector diagram for reaction A + BC → AB + C. Velocity, scattering angle, and the solid angle in the center-of-mass coordinate systems are U, θ, and Δw, respectively. The corresponding quantities in the laboratory coordinate system are V, Θ, and $\Delta\Omega$

The velocity of the two reactants in the center-of-mass coordinate system is:

$$U_A = V_A - V_{C.M.} \tag{2.3}$$

$$U_{BC} = V_{BC} - V_{C.M.} \tag{2.4}$$

After the reaction, the products AB and C in the center-of-mass coordinate system are:

$$U_{AB} = V_{AB} - V_{C.M.} \tag{2.5}$$

$$U_C = V_C - V_{C.M.} \tag{2.6}$$

As shown in Fig. 2.3, the velocity of product AB is detected as V_{AB} at laboratory angle Θ. The corresponding angle and velocity in the center-of-mass coordinate system are θ and U_{AB}. This conversion process can simply be obtained directly from vector algebra using Newton diagram.

In addition to the transformation of angle and velocity, the signal intensity must also be transformed from the laboratory coordinate system $I(\Theta, V_{AB})$ to the center-of-mass coordinate system $I(\theta, U_{AB})$. A Jacobian factor is needed for the transformation between different coordinate systems. Assuming the detector's receiving area in the experiment is ΔA, the corresponding solid angle of the scattering center is $\Delta\Omega$, and the scattering intensity of product AB with speed V_{AB} at the laboratory angle θ is $I(\Theta, V_{AB})$, and then, the received flux of scattering product AB at the laboratory angle θ is $I(\theta, V_{AB})\Delta\Omega$. As shown in Fig. 2.3, the flux should be equal to the received flux ($I(\theta, U_{AB})\Delta w$) of scattering product AB at angle θ in the center-of-mass coordinate system.

$$I(\Theta, V_{AB})\Delta\Omega = I(\theta, U_{AB})\Delta w \tag{2.7}$$

It can be seen from Fig. 2.3, and the solid angle of the detector is inversely proportional to the square of the speed, i.e.,

$$\Delta\Omega \propto \frac{\Delta A}{V_{AB}^2} \qquad (2.8)$$

$$\Delta W \propto \frac{\Delta A}{U_{AB}^2} \qquad (2.9)$$

From (2.7), (2.8), and (2.9), we can get:

$$I(\theta, U_{AB}) = J(\text{Lab} \rightarrow \text{C.M})I(\Theta, V_{AB}) \qquad (2.10)$$

in which, $J(\text{Lab} \rightarrow \text{C.M}) = \frac{U_{AB}^2}{V_{AB}^2}$ is the Jacobian factor for the transformation from the laboratory coordinate system to the center-of-mass coordinate system.

In this way, the angle, speed, and signal intensity detected in the laboratory coordinate system is transformed to that in the center-of-mass coordinate system. Thus, angular and velocity distribution of the center-of-mass coordinate system can be obtained, which enables us to carry out detailed analysis of the dynamics of the scattering process.

2.2 HRTOF Technique

Although H is the simplest atom in the universe, there is no particularly good way to detect it. Before the mid-1980s, the detection of the H atom was mostly dependent on the electron impact ionization. But this method is susceptible to the background interference of H_2 in the vacuum chamber. Yang et al. [20] improved the signal-to-noise ratio of detection by improving the vacuum in the ionizing region. However, due to the lightweight of the H atom, the velocity of product H atom from the reaction is very high; thus, the sensitivity of the measurement is enormously limited. The development of vacuum ultraviolet (VUV) generated by nonlinear medium offers scientists a very sensitive H atom detection method: $(1 + 1')$ resonance-enhanced multiphoton ionization (REMPI) [15]. In this method, firstly the H atom is excited from its ground state ($n = 1$) to ($n = 2$) state by Lyman-α transition (121.6 nm). Then, the H atom at $n = 2$ state is directly ionized by a UV light beam ($\lambda < 365$ nm) (Fig. 2.4). The produced hydrogen ion (H^+) can be detected using general ion detection methods. The sensitivity of this method is high, but due to the lightweight of a hydrogen ion, it is vulnerable to the stray electric field and the space charge effect in the surrounding space, which limits the translational energy resolution.

HRTOF technique was gradually developed in the 1980s [2–5, 16, 17, 23, 27, 29]. The technique has the advantage of both high sensitivity and high resolution. The development of the technique provides us a powerful tool to get state-resolved differential cross section at an unprecedented high level of energy resolution and detection limitation. The first step of HRTOF is the same as $(1 + 1')$ REMPI. The second step, different from REMPI, is exciting the H atom from its $n = 2$ electronic

Fig. 2.4 The energy levels
of H atom and the schematic
figure of 1 + 1′ REMPI and
Rydberg "tagging" of H
atom

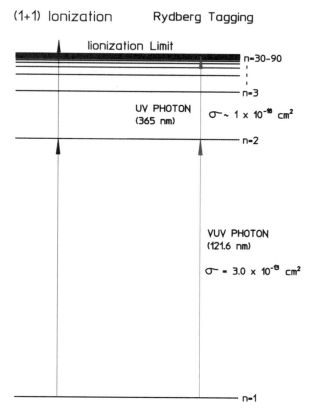

state to its high Rydberg state ($n = 30\text{–}90$) (Fig. 2.4). The high Rydberg H atoms are known to be long lived and can be field-ionized. They are field-ionized using a relatively strong electric field just before the moment reaching detector. During the flight, hydrogen atoms at the Rydberg state remain electrically neutral. The Coulomb repulsion of REMPI and the influence of the external electromagnetic field are avoided, so very high resolution can be achieved. In addition, H atom itself has no problem of the distribution of vibrational states, and the energy gap of the electronic ground state and the first electronic excited state is up to 10.2 eV. Most of the H atoms in the experiment are in the electronic ground state, so by probing their kinetic energy, the internal energy distribution of their concomitants can be easily obtained using momentum and energy conservation equations. Moreover, since the hydrogen atom is the lightest atom in nature, so compared with the other substances, a hydrogen atom has a better velocity resolution under the same energy difference. The success of this technique provides us a promising detection method for the study of elementary chemical reaction kinetics.

In early days, HRTOF technique was used to study gas-phase photodissociation whose products contain hydrogen atoms (or deuterium atoms), for example, a variety of hydrogen halides, HCN, H_2O, H_2S, NH_3, PH_3, C_2H_2, CH_3SH, CH_3NH_2, HCOOH, HFCO, HN_3 [8, 10, 37]. These processes constitute a major chemical process in interstellar chemistry, atmospheric chemistry, combustion chemistry, plasma chemistry, and other fields. HRTOF has high resolution and high signal-to-noise ratio and thus provides us a lot of detailed information to understand the mechanism of the molecular photodissociation. From the beginning of the 1990s, combined with the crossed molecular beam technique, it is used to study the elastic, inelastic, and reaction collision, including some of the most fundamental chemical reactions, for example, $H + H_2$, $H + D_2$, $O + H_2$ [11, 21, 29].

The key technique of HRTOF is to produce a laser at 121.6 nm and overlap this laser with another laser beam at 365 nm exactly both in space and in time. The laser system of the experiment is introduced here, as shown in Fig. 2.5.

We used the third harmonic of a YAG laser (Spectra Physics, Model Pro-290, 30 Hz), 355 nm (about 100 mJ), to pump a dye laser (Sirah, Model PRSC-G-30), generating 425 nm. The second harmonic 212.5 nm (about 2 mJ) is generated with a BBO crystal. The second harmonic of YAG laser, 532 nm, (approximately 400 mJ) is divided into two beams by the beam splitter, one for pumping another dye laser (Continuum, Model ND6000) to generate 845 nm (approximately 10 mJ). The two laser beams at 212.5 nm and 845 nm are overlapped both in space and in time by adjusting the optical path and then through a fused quartz lens focused in a static gas cell, made from stainless steel, and locating in the wall of the main chamber. The cell is filled approximately 80 torr Kr/Ar (1:3) mixture gas. As shown in Fig. 2.6, a 121.6 nm VUV laser is generated when 845 nm and 212.5 nm lasers pass Kr/Ar mixture gas via four-wave mixing [22]. The transmitted 532 nm after

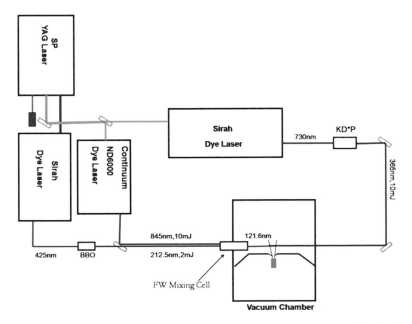

Fig. 2.5 The arrangement of the H atom Rydberg tagging laser system. Reprinted with permission from Ref. [26]. © 2005, American Institute of Physics

the beam splitter is used to pump a third dye laser (Sirah, Model PRSC-G-24), generating a 730 nm. Then, second harmonic 365 nm (approximately 10 mJ) is generated with a KD*P crystal. The 121.6 nm and 365 nm laser beams intersect the molecular beam at the reaction zone at the same time (i.e., the molecular beam, 121.6 nm laser, and 365 nm laser are overlapped exactly in time and space in the reaction zone) and excite the H atom to its high Rydberg levels ($n = 30$–90) via a two-step process, as shown in Fig. 2.4.

Tunable generation of short-wavelength VUV laser is beyond the short-wavelength limit of crystals (BBO, KD*P, etc.). Short-wavelength VUV laser is often generated through third harmonic or sum frequency generation by nonlinear effects of the inert gas or metal vapor. The 121.6 nm VUV laser we use in the first step of the hydrogen atom excitation is generated by four-wave mixing by the nonlinear effects of the inert gas Kr [22]. Its mechanism is shown in Fig. 2.6. Kr atoms in the electronic ground state $4p^6$ are excited to the excited state $4p^5 5p$ by resonance absorption of two 212.55 nm ($2\omega_R$) photons, and then induced by 845 nm (ω_T) laser to emit a same photon, and then continue the transition to ground state, thus generating 121.6 nm VUV light ($2\omega_R - \omega_T$).

The different wavelength and polarization of VUV can be obtained by adjusting the ω_T wavelength and polarization. Two factors usually affect the conversion efficiency in this process: (1) the absorption of the generated VUV in the medium and (2) phase matching. The water vapor existing in the medium will absorb the vacuum ultraviolet light, so before the gas is filled into the four-wave mixing cell, we firstly use a mixture of dry ice and alcohol to freeze impurities like water in the gas. In order to optimize the phase matching of the medium, we mix Kr gas with a certain amount of Ar gas. For more information on vacuum ultraviolet laser produced by nonlinear optical techniques, see Ref. [24].

Fig. 2.6 Generation of 121.6 nm VUV laser

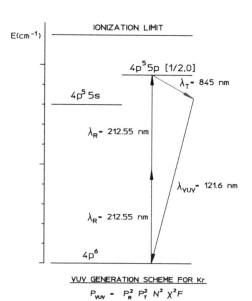

2.3 Experimental Apparatus

2.3.1 Vacuum System

Vacuum system of the HRTOF crossed molecular beam apparatus consists of three parts, as shown in Fig. 2.7: source I chamber, source II chamber and the main chamber. Two source chambers are connected to the main chamber through respective skimmer, and the two source chambers are completely isolated. The drawings of vertical and horizontal cut view of the chamber are shown in Figs. 2.8 and 2.9.

Three chambers are pumped independently. The two source chambers are pumped by a maglev turbo molecular pump (BOC EDWARDS, Model STP-A2203C) with a pumping speed of 2000 l/s. The main chamber is pumped by the combination of a maglev turbo molecular pump (same as in source chamber) and a cryopump (Austin Scientific, Model M125). When the general valves are not working, the vacuum background in all the three chambers can achieve 10^{-8} torr; when the general valves are working, the vacuum in the source I chamber is about 10^{-6} torr and the vacuum in both the source II chamber and the main chamber is about 10^{-5} torr.

2.3.2 Molecular Beam Source

In crossed molecular beam experiments, we can improve the beam intensity using supersonic molecular beam. It is necessary to consider how to obtain high-resolution time-of-flight spectra. As described previously, a supersonic molecular beam has a narrow speed distribution. But after all there is a speed distribution.

Fig. 2.7 Photograph of HRTOF crossed molecular beam machine

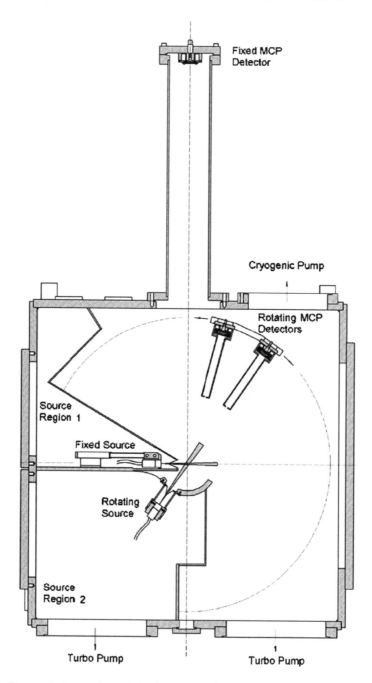

Fig. 2.8 The vertical cut view of the instrument that shows the arrangement of the main components of the instrument. Reprinted with permission from Ref. [26]. © 2005, American Institute of Physics

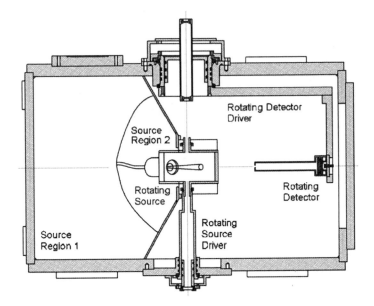

Fig. 2.9 The horizontal cut view of the instrument that shows the driving mechanisms for the rotating detector and the rotating source. Reprinted with permission from Ref. [26]. © 2005, American Institute of Physics

Fig. 2.10 Schematic of resolution limited by speed distribution of molecular beams in crossed molecular beam experiment

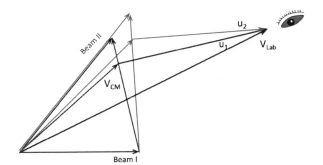

Figure 2.10 is a schematic of resolution limited by speed distribution of molecular beams in crossed molecular beam experiment. We can see that molecular speed distribution leads to the collision energy distribution. The same laboratory speed of a product in laboratory coordinate system corresponds to a product from different collision energy, different speed, and different scattering angle. The final resolution is not attributed to the broadening of collision energy, caused by the speed distribution of the molecular beam, but to the broadening of the center-of-mass velocity (V_{CM}), caused by speed distribution of the molecular beam as well. There are different experimental resolutions at different detection angles. For example, the resolution is high at the perpendicular direction to the V_{CM}, and it is low at the

parallel direction of V_{CM}. After we further analyzed, we also found that speed distribution of heavy molecules has greater impact if two molecular beams have the same speed distribution. Therefore, it is more important to optimize the molecular beam of a heavy molecule under the same conditions. Secondly, since the molecular beams have a divergence in space, and the crossing angle of the two molecular beams has a distribution, the resolution is reduced as well.

A fixed general valve is installed in the source I chamber. H_2 (or HD, D_2) molecules are collimated by a skimmer and then collide and react with fluorine atom beam, which is double-skimmer collimated, in scattering chamber. The two molecular beam sources in this experiment will be described below.

2.3.2.1 Source I Chamber

In order to obtain a low speed to meet low collision energy requirements of the experiment and single-state molecular beam to study state-to-state dynamics, we modify the general valves to enable them to work at liquid nitrogen temperature (~ 78 K). To ensure the general valves working at liquid nitrogen temperature, the adjustment of the valves is also at liquid nitrogen temperature. The relative position of the valve body and the valve head is adjusted at liquid nitrogen temperature, and then, they are sealed with a vacuum seal (Varian, Torr Seal). The entire valve is fixed to an oxygen-free copper base, which is fixed to the source I chamber. During the experiment, the copper base is cooled by liquid nitrogen. The high thermal conductivity of oxygen-free copper allows the valve to be well cooled.

In the experiment, the back pressure of H_2 (or HD, D_2) gas source is 2 atm. The speed of 99.999 % pure H_2, HD, and D_2 (HSG) after supersonic expansion is about 1.38, 1.24, and 1.03 km/s, respectively. The H atoms are generated by vacuum gauge ionization from H_2 and then blown by the molecular beam into the scattering chamber. The molecular beam speed is determined using the Rydberg tagging time-of-flight technique.

Under normal circumstances, normal-H_2 consists of para-H_2 and ortho-H_2 with the ratio 1:3. The para-H_2 has antisymmetric nuclear spin configuration, corresponding only to the energy level at even-number rotational quantum states ($j = 0$, 2, 4,...); ortho-H_2 has symmetrical nuclear spin configuration, corresponding only to the odd-number rotational quantum states ($j = 1, 3, 5,...$). This is because the H atom nuclear spin quantum number $I = 1/2$ is fermion; thus, the nuclear exchange symmetry of the total wave function is antisymmetric. According to the transition selection rule, the transition between para-H_2 and ortho-H_2 is forbidden, so at low temperature, the para-H_2 molecules, originally at even-number j energy level, are populated to the lowest state, $v = 0, j = 0$; the ortho-H_2 molecules, originally at odd-number j energy level, are populated to the lowest state $v = 0, j = 1$. In fact, the selection rule is not absolute. Due to the weak interactions between the nuclear spins and electron, collision between molecules can induce the transition between para- and ortho-H_2, but the balancing process is very long, often takes several months, until all the molecules at $j = 1$ energy level transit to $j = 0$ energy level. If adding a small amount of paramagnetic substances as a catalyst at low temperature, this process will be greatly shortened. The molecules at $j = 1$ energy level can soon transit to $j = 0$ energy level. At this time, heating these H_2 at $j = 0$ energy level in the absence of the catalyst, we can get the para-H_2 ($j = 0, 2, 4, \ldots$). The obtained para-H_2 can be maintained for several weeks (a container from magnetic materials must be avoided; we use all-aluminum cylinder storage). In the experiment, we passed the H_2 slowly through a catalyst (IONEX, ortho-para conversion catalysis) at the temperature lower than 30 K (cryopump, Austin Scientific, Model M600) to obtain the para-H_2 [25].

Same as hydrogen, under normal circumstances, deuterium (normal-D_2) consists of ortho-D_2 and para-D_2. However, because the D atom nuclear spin quantum number $I = 1$ is boson, nuclear exchange symmetry of the total wave function is symmetrical. Para-D_2 corresponds to the energy levels at even-number rotational quantum states ($j = 0, 2, 4, \ldots$); ortho-D_2 corresponds to odd-number states ($j = 1, 3, 5, \ldots$). The ratio of ortho-D_2 to para-D_2 is 2:1. We prepared ortho-D_2 using the same method as para-H_2.

HD is not homonuclear diatomic molecule, so there is no problem of nuclear exchange symmetry.

The purity of the prepared para-H_2 (or ortho-D_2) is measured by Raman spectroscopy. Figure 2.11 is the Raman spectra of n-H_2 and the p-H_2 prepared by the above method, measured at room temperature. The purity of p-H_2 can be obtained by using the thermal statistical distribution of n-H_2 at room temperature and normalizing the Raman scattering cross section and the detection efficiency. Prepared by the above method, the purity of the p-H_2 is higher than 96 %, and the purity of o-D_2 is higher than 90 %.

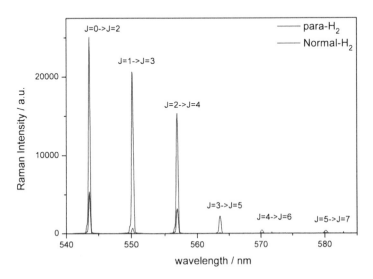

Fig. 2.11 Raman spectra of para-H_2 and normal-H_2

In the experiment, in order to study the impact of the rotation of hydrogen (or deuterium) on the dynamics, we use two gas samples: n-H_2 and p-H_2 (or n-D_2 and o-D_2). Taking H_2 as an example, H_2 molecular beam is generated by H_2 through cold valve supersonic expansion, so the beam speed ratio $v/\Delta v$ can reach about 30, and all molecules of the beam are populated at the lowest rotational energy level. Considering that the purity of the prepared p-H_2 is 96 %, H_2 is at the $j = 0$ and $j = 1$ rotational energy levels to a ratio of 96:4 in the p-H_2 molecular beam, while H_2 is at the $j = 0$ and $j = 1$ rotational energy levels to a ratio of 1:3 in the n-H_2 molecular beam. In order to reduce the system error of the experiment, in each detection angle, data are acquired under the same experimental conditions using the two gases in turn more than 10 rounds. The pressure of both gas sources is controlled as the same, so the molecule densities of the two molecular beams are equal, whereby we can obtain H atom TOF spectra of the pure F + $H_2(j = 0)$ and F + $H_2(j = 1)$ reaction by acquiring F + n-H_2 and F + p-H_2 generated H atom TOF spectra, following mathematical transformation derivation:

The TOF spectrum signal of F + n-H_2 and F + p-H_2 is represented by the TOF spectra signal of F + $H_2(j = 0)$ and F + $H_2(j = 1)$ as

$$TOFS(p-H_2) = 96\% \times TOFS(j = 0) + 4\% \times TOFS(j = 1) \qquad (2.11)$$

$$TOFS(n-H_2) = 1/4 \times TOFS(j = 0) + 3/4 \times TOFS(j = 1) \qquad (2.12)$$

From (3.1) and (3.2), it can be deduced,

$$TOFS(j = 0) = 1.056 \times TOFS(p - H_2) - 0.056 \times TOFS(n - H_2) \qquad (2.13)$$

$$TOFS(j = 1) = 1.344 \times TOFS(n - H_2) - 0.344 \times TOFS(p - H_2) \qquad (2.14)$$

2.3.2.2 Source II Chamber

In the beam source II chamber, the discharge fluorine atom beam source is installed in a rotational mechanical device, which can be rotated around the reaction center (i.e., the intersection of the two molecular beams) (see Fig. 2.8); thus, the collision energy is easily changed by changing the crossing angle of two molecular beams. The adjustment range of molecular beam crossing angle is 42.5–135° in steps of 2.5°.

At the beginning, we used double-stage pulsed discharge, the pre-ionization, and the main discharge, but either time delay or pulse width of the pre-ionization is not controllable. Additionally, the two discharges are in the same space. Figure 2.12 is the TOF of F + H_2 cross-beam experiment in this way, in which rotational states of HF are barely resolvable. The resolution can be improved by reducing the pulse width of the main discharge, but the signal also decreases rapidly. The reason for the improvement in the resolution is that when the discharge pulse width is narrowed, the beam source temperature is low and the F atom beam generated by supersonic expansion process is cooled to a lower temperature and has a relatively

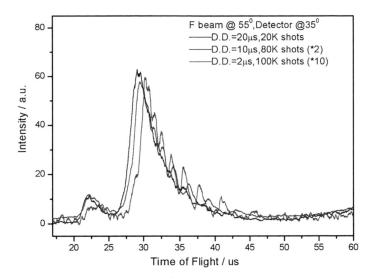

Fig. 2.12 Results of crossed beam experiment with double-stage discharge fluorine atom source, but either time or pulse width of both is not controllable, and the two discharges are in the same space

high speed ratio $v/\Delta v$. There is a certain distance from the general valve to crossed beam collision area, so it is easier to separate F atoms at different speeds when the discharge pulse width is narrowed. However, the F atom beam intensity becomes lower at the same time, resulting in low signals. In addition, the discharge width also affects the speed of the F atoms. The wider the discharge width, the higher the temperature of the beam source, and the faster the F atom beam.

Based on the above experiment and analysis, we conclude that in order to obtain a high resolution, the discharge pulse width must be shortened effectively, the (translational) temperature of the molecular beam must be lowered to get a high speed ratio, and the F atoms with a different speed must be better separated in space. However, it is necessary to improve the efficiency of the discharge generation of the F atoms, so as to complete the experiments of the small cross section. Improved double-stage pulsed discharge device is shown in Fig. 2.13 [28]. The double-stage pulsed discharge device consists of the general valve (⑧), three stainless electrodes (①③⑤), and three ceramic insulators (②④⑥) (Fig. 2.13). The electrodes and the insulators are concentrically fixed in a PTFE insulator holder (⑦). In order to prevent discharge between the electrodes and the general valve, the ceramic insulator ⑥ (thickness 2 mm) is twice thicker than the other two ceramic insulators ②④ (thickness 1 mm). For improving the discharge efficiency, the center hole diameter of the ceramic insulators ②④ is 2 mm, which is larger than the center hole diameter of the other insulators (1 mm). After the general valve is opened, the gas molecules pass through the ceramic insulator ⑥ and the electrode plate ⑤, reach the predischarge zone, and predischarge under the pulsed electric field (~ 10 μs) between the electrode plates ③ and ⑤. After about 10 μs

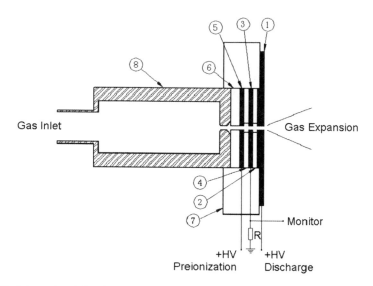

Fig. 2.13 Schematic of fluorine atom beam source by double-stage pulsed discharge method. Reprinted with permission from Ref. [28]. © 2006, American Institute of Physics

the main discharge under the pulsed electric field (~ 1 μs) between the electrode plates ① and ③. The voltage applied to the electrode plates is adjusted in accordance with the gas sample, typically between 800 and 1200 V. The predischarge current can be adjusted by an adjustable resistor in the circuit, to ensure a stabilized main discharge. Intermediate electrode plate is grounded through the resistor R. The real-time condition of the discharge is monitored by an oscilloscope. Circuit diagram is shown in Fig. 2.14. The key point of this discharge mode is the separation of the predischarge and main discharge in both time and space, so that when the pulsed electric field of the main discharge is open, most of the predischarge generated ions, electrons, and metastable atoms exactly reach the main discharge zone, which greatly enhance the efficiency and stability of the main discharge. The time sequence of double-stage pulsed discharge and monitoring of current of pulsed discharge is shown in Fig. 2.15. Experiments show that this double-stage pulsed discharge mode greatly improves the intensity and the speed ratio ($v/\Delta v \sim 10$) of the fluorine atom beam, so measuring state-to-state differential cross section with high resolution under low collision energy becomes possible.

2.3.3 Scattering Chamber

The main chamber is the largest in the three chambers, in which reaction and detection of the product are carried out. The main chamber is equipped with two rotating detectors, which can rotate in the plane determined by both molecular

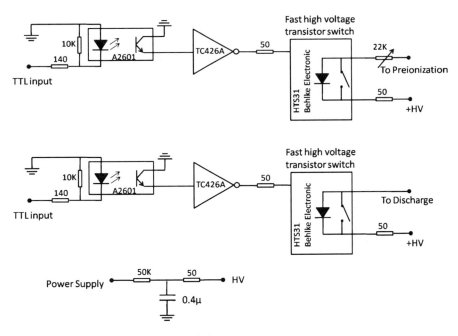

Fig. 2.14   Circuit of double-stage pulsed discharge

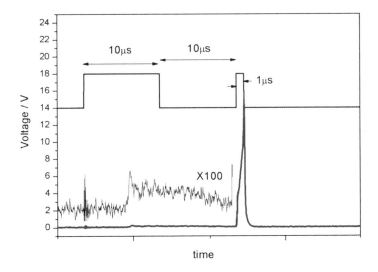

Fig. 2.15   Time sequence and currents of pulsed discharge monitored by an oscilloscope

beams, around the reaction center with a rotational range of 220°, covering all angles from the forward scattering to the backward scattering of the product, as shown in Fig. 2.9. The two rotation detectors are fixed with a 20° angle and can be used simultaneously, so that we can save collection time and improve the experimental efficiency and reliability. Vertical upward direction in the reaction center is also equipped with a fixed detector (this detector is replaced by vertically movable electrode plates for the fine tuning of the optical path in cross-beam experiments) with the flight path of 1 m, which can be used to study the photo-dissociation dynamics with higher resolution.

2.3.4 Detection and Acquirement System

The detector is a microchannel plate (MCP) Z-stack detector. The incident surface of the first MCP and the outgoing surface of the third MCP are applied with high voltage DC, about −2700 and −270 V, respectively. A grid of high transmittance (90 %) is grounded about 5 mm at the front of the first MCP. In the experiment, the cross-beam produced H (or D) atoms are two-step resonance excited to high Rydberg state, and then reach the grid after approximately 318 mm free flight. The Rydberg tagging H atoms are then immediately field-ionized by the strong electric filed applied between the front plate of the z-stack MCP detector and the grid. The generated hydrogen atom ions (H^+) are accelerated by the electric field and fly to the MCP. The signal of hydrogen ions (H^+) is amplified by six to seven orders of magnitude by the MCP. The current output of the MCP is received by an anode and amplified 20 times by a preamplifier (ORTEC, Model VT120). The detected signal is discriminated and shaped by a discriminator (ORTEC, Model 935) and finally counted by a multichannel scaler (FAST ComTec, Model P7888) and recorded by a computer, as shown in Fig. 2.16.

Hydrogen ions (H^+), generated at the same time of "tagging" H atom at Rydberg state, can seriously affects the signal-to-noise ratio of the experiment. So we add a weak electric field (20 V/cm) perpendicular to the direction of flight in front of the flight zone for the removal of ions. The presence of the weak electric field does not ionize the high Rydberg state H atoms. Conversely, it is able to increase their lifetimes [29]. When the angular momentum quantum number is relatively small, the lifetime of the Rydberg atoms is proportional to n^3; when the angular momentum quantum number is relatively large, the lifetime is proportional to n^5. For example, when $n = 40$, the lifetimes of $l = 0$, 1, and 2 Rydberg state hydrogen atoms are 111, 12, and 34 μs, respectively. When l increases to 39, the

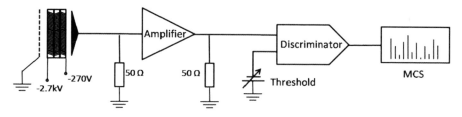

Fig. 2.16 Schematic of single detection and acquirement system

lifetime is 9.32 ms. According to dipole transition rule of atomic angular momentum $\Delta l = \pm 1$, l of Rydberg state atom by a two-step transition from the ground state $n = 1$, equals to a maximum of 2. At the presence of an electric field, the dipole transition rules is replaced by the magnetic quantum number selection rule $\Delta m = \pm 1$; thus, it is possible to get a high l value by a two-step transition.

The following physical processes may occur during H atom Rydberg tagging and flight [29]: (a) the Rydberg state hydrogen atoms emit spontaneously and transit to low n value energy level; (b) the Rydberg state hydrogen atoms transit to a low or high Rydberg state, or even ionized, by stimulated emission or absorption by electromagnetic radiation field in the chamber; (c) the Rydberg state hydrogen atoms collide with other atoms or molecules in the chamber, which come from molecular beams or from background gas. The collisions result in a change in quantum state of the Rydberg state hydrogen atoms, even ionization; (d) change in the quantum state due to the presence of an electric field. It is difficult to assess the combined effect of various processes; especially for many processes, we do not know their reaction cross sections. Photodecomposition of NH_3 and CH_4 is a good example. Experiments show that [29] between $n = 30$ and $n = 60$, there is no significant change in hydrogen atom TOF spectra within the flight time up to at least 500 μs. For our experiments, the measured Rydberg state hydrogen atom TOF spectra can truly reflect the initial speed distribution of the hydrogen atoms generated by the crossed beam reaction.

2.3.5 The Instrument Resolution

To determine the resolution of an experimental apparatus, the photodissociation of the diatomic molecules is a good metric. Photodissociation products of diatomic molecules are two atoms without rovibrational structure. The energy levels of the atoms are discrete, and the energy levels are very narrow, so it is easy to know the electronic states of the atom products. Usually, relatively narrow energy levels do not result in widening of the product translational energy spectra, which is beneficial for real instrument resolution. We determine the resolution using photodissociation of hydrogen iodide (HI), at 121.6 nm. At 121.6 nm excitation, there are two main channels of HI photodissociation:

$$HI + h\nu \rightarrow H(n = 1) + I\left({}^2P_{3/2}\right)$$

$$HI + h\nu \rightarrow H(n = 1) + I^*\left({}^2P_{1/2}\right)$$

Figure 2.17 is the TOF spectra of H atom from HI photodissociation at 121.6 nm. When the flight distance is about 1 m, the time resolution is up to 0.06 %, corresponding to a translational energy resolution of 0.12 %. Such a high resolution makes quantum state resolution of photodissociation possible. The results on crossed beam experiments can be found in the next two chapters.

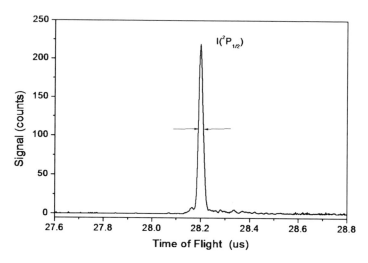

Fig. 2.17 TOF spectra of H atom from HI photodissociation at 121.6 nm. Reprinted with permission from Ref. [26]. © 2005, American Institute of Physics

References

1. Althorpe SC, Fernandez-Alonso F, Bean BD et al (2002) Observation and interpretation of a time-delayed mechanism in the hydrogen exchange reaction. Nature 416:67–70
2. Ashfold MNR, Bennett CL, Dixon RN (1986) Dissociation dynamics of NH$_3$(Ã^{1}A$_2$): experiment and theory. Faraday Discuss Chem Soc 82:163–175
3. Ashfold MNR, Dixon RN, Kono M et al (1997) Near ultraviolet photolysis of ammonia and methylamine studied by H Rydberg atom photofragment translational spectroscopy. Philos Trans R Soc Lond a-Math Phys Eng Sci 355:1659–1674
4. Biesner J, Schnieder L, Schmeer J et al (1988) State selective photodissociation dynamics of A-tilde state ammonia. I. J Chem Phys 88:3607–3616
5. Biesner J, Schnieder L, Ahlers G et al (1989) State selective photodissociation dynamics of A-tilde state ammonia. II. J Chem Phys 91:2901–2911
6. Casavecchia P, Capozza G, Segoloni E et al (2005) Dynamics of the O(^{3}P) + C$_2$H$_4$ reaction: Identification of five primary product channels (vinoxy, acetyl, methyl, methylene, and ketene) and branching ratios by the crossed molecular beam technique with soft electron ionization. J Phys Chem A 109:3527–3530
7. Dai DX, Wang CC, Harich SA et al (2003) Interference of quantized transition-state pathways in the H + D$_2$ → D + HD chemical reaction. Science 300:1730–1734
8. Dixon RN, Hwang DW, Yang XF et al (1999) Chemical "double slits": dynamical interference of photodissociation pathways in water. Science 285:1249–1253
9. Even U, Jortner J, Noy D et al (2000) Cooling of large molecules below 1 K and He clusters formation. J Chem Phys 112:8068–8071
10. Harich SA, Hwang DWH, Yang XF et al (2000) Photodissociation of H$_2$O at 121.6 nm: a state-to-state dynamical picture. J Chem Phys 113:10073–10090
11. Harich SA, Dai DX, Wang CC et al (2002) Forward scattering due to slow-down of the intermediate in the H + HD → D + H$_2$ reaction. Nature 419:281–284
12. Herschbach DR (1986) Molecular dynamics of elementary chemical reactions. In: Nobel Lecture

13. Kantrowitz A, Grey J (1951) A high intensity source for the molecular beam. Part I Theoretical. Rev Sci Instrum 22:328–332
14. Kistiakowsky GB, Slichter WP (1951) A high intensity source for the molecular beam. Part II. Experimental. Rev Sci Instrum 22:333–337
15. Krautwald HJ, Schnieder L, Welge KH et al (1986) Hydrogen-atom photofragment spectroscopy: photodissociation dynamics of H_2O in the B-X absorption band. Faraday Discuss Chem Soc 82:99–110
16. Langford SR, Batten AD, Kono M et al (1997) Near-UV photodissociation dynamics of formic acid. J Chem Soc, Faraday Trans 93:3757–3764
17. Langford SR, Regan PM, Orr-Ewing AJ et al (1998) On the UV photodissociation dynamics of hydrogen iodide. Chem Phys 231:245–260
18. Lee YT (1987) Molecular beam studies of elementary chemical processes. Science 236:793
19. Lee YT, Mcdonald JD, Lebreton PR et al (1969) Molecular beam reactive scattering apparatus with electron bombardment detector. Rev Sci Instrum 40:1402–1408
20. Lin JJ, Hwang DW, Harich S et al (1998) New low background crossed molecular beam apparatus: low background detection of H_2. Rev Sci Instrum 69:1642–1646
21. Liu XH, Lin JJ, Harich S et al (2000) A quantum state-resolved insertion reaction: $O(^1D) + H_2(J = 0 \rightarrow OH(^2\prod, v, N) + H(^2S)$. Science 289:1536–1538
22. Marangos JP, Shen N, Al HME (1990) Broadly tunable vacuum-ultraviolet radiation source employing resonant enhanced sum-frequency mixing in krypton. J Opt Soc Am B 7:1254–1259
23. Mordaunt DH, Ashfold MNR, Dixon RN et al (1998) Near threshold photodissociation of acetylene. J Chem Phys 108:519–526
24. Puell H, Spanner K, Falkenstein W et al (1976) Third-harmonic generation of mode-locked Nd: glass laser pulses in phase-matched Rb-Xe mixtures. Phys Rev A 14:2240
25. Qiu M (2006) thesis: high resolution crossed molecular beams study on the F + H_2 reaction. In: Dalian Institute of Chemical Physics, CAS. Dalian, China
26. Qiu MH, Che L, Ren ZF et al (2005) High resolution time-of-flight spectrometer for crossed molecular beam study of elementary chemical reactions. Rev Sci Instrum 76:083107
27. Reed CL, Kono M, Langford SR et al (1997) Ultraviolet photodissociation dynamics of formyl fluoride.2. Energy disposal in the H + FCO product channel. J Chem Soc, Faraday Trans 93:2721–2729
28. Ren ZF, Qiu MH, Che L et al (2006) A double-stage pulsed discharge fluorine atom beam source. Rev Sci Instrum 77:016102
29. Schnieder L, Seekamp-Rahn K, Wrede E et al (1997) Experimental determination of quantum state resolved differential cross sections for the hydrogen exchange reaction H + D_2 → HD + D. J Chem Phys 107:6175–6195
30. Scoles G (ed) (1988) Atomic and molecular beam methods, vol 1. Oxford University Press, Oxford
31. Scoles G (ed) (1992) Atomic and molecular beam methods, vol 2. Oxford University Press, Oxford
32. Smalley RE, Wharton L, Levy DH (1977) Molecular optical spectroscopy with supersonic beam and jets. Acc Chem Res 10:139–146
33. Taylor EH, Datz S (1955) Study of chemical reaction mechanisms with molecular beams. The reaction of K with HBr. J Chem Phys 23:1711–1718
34. Townsend D, Lahankar SA, Lee SK et al (2004) The roaming atom: straying from the reaction path in formaldehyde decomposition. Science 306:1158–1161
35. Wang J, Shamamian VA, Thomas BR et al (1988) Speed ratios greater than 1000 and temperature less than 1-mK in a pulsed He beam. Phys Rev Lett 60:696–699
36. Yan S, Wu YT, Zhang B et al (2007) Do vibrational excitations of CHD_3 preferentially promote reactivity toward the chlorine atom? Science 316:1723–1726
37. Yang XF, Hwang DW, Lin JJ et al (2000) Dissociation dynamics of the water molecule on the (A) over-tilde B-1(1) electronic surface. J Chem Phys 113:10597–10604

38. Yin HM, Kable SH, Zhang X et al (2006) Signatures of H_2CO photodissociation from two electronic states. Science 311:1443–1446
39. Yuan KJ, Cheng Y, Cheng L et al (2008) Nonadiabatic dissociation dynamics in H_2O: competition between rotationally and nonrotationally mediated pathways. Proc Natl Acad Sci USA 105:19148–19153
40 Yang, X, Lin J, Lee YT, Blank DA, Suits AG, and Wodtke AM (1997) Universal crossed molecular beams apparatus with synchrotron photoionization mass spectrometric product detection. Rev Sci Instrum. 68(9): 3317–3326.

Chapter 3
Dynamical Resonances in F + H$_2$ Reactions

F + H$_2$ reaction first attracted attention due to the application of chemical laser. This is the first reaction which has product vibrational state resolved measurements. Using chemical laser [20] and infrared light emitting [21, 22], researchers found that the population of the product HF vibrational states is highly inverted. Crossed molecular beam studies of this system are the main work of Yuan Tseh Lee's Nobel Prize in Chemistry in 1986 [12, 13]. In this chapter, studies on resonance phenomenon in the F + H$_2$ reaction are mainly described. In Sect. 3.1, studies on resonance in the F + H$_2$ reaction are reviewed; crossed molecular beam studies in the F(^{2}P$_{2/3}$) + H$_2$ → HF + H reaction are introduced in Sects. 3.2 and 3.3 discusses the studies of the F(^{2}P$_{2/3}$) + HD → HF + H reaction, and the last section is a summary.

3.1 Review of Reaction Resonances in F + H$_2$ Reaction

The reaction resonance in the F + H$_2$ reaction and the isotope substituted reactions has been a central topic in the study of chemical reaction dynamics in the last few decades [1, 2, 4, 15]. Since the first theoretical prediction of reaction resonances in the F + H$_2$ reaction in the early 1970s made by Schatz et al. [28] and Wu et al. [35], the search for evidence of such resonances in this reaction has attracted much attention from many top research groups in this field. In 1984, Neumark et al. [17, 18] performed a landmark crossed beam experiment on the F + H$_2$ reaction using a universal crossed molecular beams apparatus. A clear forward-scattering peak was observed for the HF($v' = 3$) product, which was assigned to reaction resonances in this reaction. Furthermore, forward scattering for the DF($v' = 4$) product from F + D$_2$ as well as the HF($v' = 3$) product from F + HD were observed, which are consistent with the F + H$_2$ experiment [19]. However, the full quantum mechanical (QM) scattering calculations of the F + H$_2$ reaction on the Stark-Werner PES (SW-PES) [31] do not support this resonance conjecture [5]. Quasi-classical trajectory (QCT) calculations on the same surface performed by Aoiz et al. [3] also exhibit forward scattering of HF($v' = 3$) in the

Z. Ren, *State-to-State Dynamical Research in the F+H$_2$*
Reaction System, Springer Theses, DOI: 10.1007/978-3-642-39756-1_3,
© Springer-Verlag Berlin Heidelberg 2014

same reaction. The SW-PES is reasonably accurate in describing the transition state (TS) region for the F + H$_2$ reaction as revealed in the negative ion photo-detachment study of the FH$_2^-$ system [16]. Observation of forward scattered HF($v' = 3$) products from QCT calculations based on classical mechanics on the SW-PES implies that the HF($v' = 3$) forward scattering observed in the experiment might be due to mechanisms other than resonances, because resonance is a quantum phenomenon, which cannot be described properly by the classical way.

In a more recent crossed beam experiment, Liu and Skodje et al. unambiguously observed a step in the total excitation function at about 0.5 kcal/mol collision energy in the F + HD → HF + D reaction [29]. Theoretical analysis based on the SW-PES attributed this step to a single reaction resonance in the F + HD reaction. This is the first conclusive evidence for the reaction resonance in the F + HD reaction. Liu et al. also measured series of differential cross sections in the collision energy range of 0.4–1.18 kcal/mol [11] as well as 1.3–4.53 kcal/mol [10]. Results also show that there is a reaction resonance in this reaction. However, there are still considerable differences between theory and experiment. Although there is also a step of calculated reaction section, Fig. 3.1 shows the step height is about twice the experimental result and about 0.1 kcal/mol higher than the experiment. Moreover, when the spin–orbit effects are taken into account based on SW potential energy surface, theoretical and experimental differences are greater. Hayes made a correction to SW potential energy surface in the outlet channel and got SWMHS potential energy surface [8]. He found that the reaction resonance on this potential energy surface is substantially different from that on the SW potential energy surface. Theoretical study on the SW-PES also found that resonance in the F + HD reaction has a profound effect on the reaction rate constant at collision energies below and above the reaction barrier [33]. No step in the excitation function similar to that for the F + HD reaction was observed for the F + H$_2$ reaction [7], suggesting the dynamics for these two systems are considerably different. Whether a reaction resonance exists in the F + H$_2$ reaction still remains unclear.

Fig. 3.1 Excitation function of reaction
F + HD → HF + D.
Reprinted figure with permission from [29]. © 2000 by American Physical Society

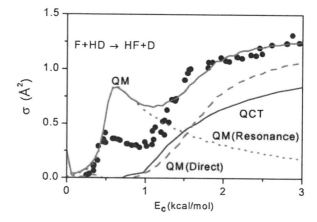

In this chapter, we used crossed molecular beam and hydrogen atom Rydberg tagging time-of-flight (TOF) technique to study the reaction F + H$_2$ → HF + H and F + HD → HF + D. With cooperation with Daiqian Xie and Donghui Zhang, who worked on calculating new potential energy surface and dynamics, we systematically studied the reaction resonance in this benchmark system.

3.2 Study on the Reaction Resonance in $F(^2P_{3/2})$ + H$_2$ → HF + H

3.2.1 Experimental Method

In this experiment, a fluorine atom beam is generated by a double-stage pulsed discharge method [27], as previously described in detail. The discharge gas is a mixture gas of 5 % F$_2$ and 95 % He. The pulse valve back pressure is about 6 atm. The fluorine atom beam contains both spin–orbit ground state $F(^2P_{3/2})$ and spin–orbit excited state $F^*(^2P_{1/2})$, and the $F(^2P_{3/2})$:$F^*(^2P_{1/2})$ ratio of relative concentrations measured by single photon autoionization is 10.7 [6]. H$_2$ molecular beam is generated by H$_2$ gas (back pressure 2 atm) through supersonic expansion with a general valve cooled by liquid nitrogen ($\sim$ 78 K). After supersonic cooling, almost all of the H$_2$ are at lowest states ($j = 0$ for p-H$_2$, $j = 0$, 1 for n-H$_2$), molecular speed ratio is very narrow ($v/\Delta v \approx 28$). Thus, there is only one state for each of the two reactants (F* reactivity is very low), so that we can achieve the state-to-state reaction, and the experimental resolution is also improved. Further, since the H$_2$ beam source is cooled, the molecular beam density in the valve is much higher than valve at room temperature within the same back pressure.

3.2.2 Experimental Results and Analysis

3.2.2.1 Experimental Observation of Reaction Resonance in $F(^2P_{2/3})$ + H$_2(j = 0)$ → HF + H

The collision angle of $F(^2P_{2/3})$ and p-H$_2$ molecular beams is 55°, while the collision energy is 0.52 kcal/mol. TOF spectra were measured at $\approx 10°$ intervals in the laboratory coordinate system. TOF spectra at three laboratory angles are shown in Fig. 3.2: $-60°$, 25°, and 95°, which correspond roughly to the forward-, sideways-, and backward-scattering directions for the HF($v' = 2$) product in the center-of-mass frame, respectively. The big structures correspond to different vibrational states of HF, while the small peaks correspond to each rotational state. The upper right corner of Fig. 3.2b displays a strong HF($v' = 3$) signal. It can be seen from the spectra that the experimental resolution is very high, almost every

Fig. 3.2 TOF spectra of the H atom product from the F + H₂(j = 0) reaction at the collision energy of 0.52 kcal/mol. TOF spectra at three laboratory angles are shown **a.** $\Theta_L = -60°$ **b.** $\Theta_L = 25°$ **c.** $\Theta_L = 95°$, which correspond roughly to the forward-, sideways-, and backward-scattering directions for the HF(v' = 2) product in the center-of-mass frame, respectively. From [24], reprinted with permission from AAAS

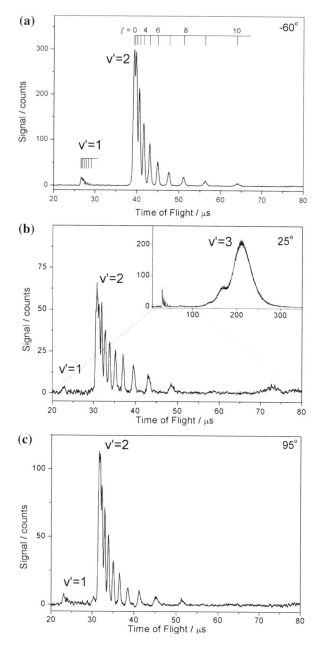

ro-vibrational state is resolved except HF(v' = 2, j = 0, 1). It can also been found that the product HF(v' = 2) has an obvious forward-scattering peak, which is the first time that a forward-scattering products of HF(v' = 2) are observed in the reaction F($^2P_{2/3}$) + H₂.

Using our self-developed software, we simulated and fitted the TOF spectra to get the distribution of products at the different speeds and different angles in the center-of-mass coordinate system, and thus, a complete differential cross section of this reaction is acquired. The software is mainly based on the conversion between the laboratory coordinate system and the center-of-mass coordinate system, as previously described in Chap. 2. By adjusting the flight distance and molecular speeds of the two molecule beams in a small range, the spectrum of the $HF(v' = 2)$ product can be fitted very well, but that of the $HF(v' = 3)$ product cannot. The main reason is that this collision energy is close to the channel's threshold of the reaction $HF(v' = 3)$ [34], and F atoms with high speed in the F atomic beam react with H_2, producing $HF(v' = 3)$.

Due to the very high resolution of the experimental data, the fitting also requires two important accurate data: the reaction energy and the HF spectrum data. The accurate reaction energy can be obtained from the accurate dissociation energy (D_0) [9] of HF, and the accurate dissociation energy (D_0) [32] of H_2, HD, and D_2. The accurate dissociation energy (D_0) of HF is from the threshold ion-pair production spectroscopy (TIPPS) experimental data [9], the accurate ionization of H (or D) atoms, and accurate electron affinity of F atoms. The reaction energy of three isotopes is summarized as follows:

$$F(^2P_{2/3}) + H_2(j = 0) \rightarrow HF + H + 11192.7 \, cm^{-1} \tag{3.1}$$

$$F(^2P_{2/3}) + HD \rightarrow HF + D + 10905.0 \, cm^{-1} \tag{3.2}$$

$$F(^2P_{2/3}) + D_2(j = 0) \rightarrow DF + D + 11110.4 \, cm^{-1} \tag{3.3}$$

HF spectral data (especially $v = 2$ high rotational and $v = 3$ data) are obtained by Dunham parameter calculation, which is fitted from the emission spectrum, so every number of the HF ro-vibrational state energy is also very accurate [14].

Daiqian Xie and Donghui Zhang et al. built a new potential energy surface (XXZ-PES) using ab initio method [36] and calculated the dynamics of $F + H_2(j = 0)$ reaction at collision energy of 0.52 kcal/mol using ABC code. Figure 3.3 shows experimental (a) and theoretical (b) 3D contour plots of this reaction, and these two data agree very well. We also compared the experimental and theoretical results at other collision energies, and the agreement is also generally good [24].

We also carried out a careful measurement of the differential cross section versus collision energy for the $HF(v' = 2)$ product in the forward-scattering direction as shown in Fig. 3.4. During the experiment, we changed the collision energy by changing the collision angle (every 2.5°). The detector was fixed in the laboratory angle corresponding to the $HF(v' = 2)$ forward-scattering direction. The measurement was repeated 10 times to reduce experimental system error. The hollow circles are the experimental data. The data in the collision energy range from 0.2 to 0.9 kcal/mol show a peak for the forward-scattering $HF(v' = 2)$ product at the collision energy of 0.52 kcal/mol. Theoretical calculation on XXZ

Fig. 3.3 Experimental
(**a**) and theoretical (**b**) 3D
contour plots for the product
HF translational energy and
angle distributions for the
F + H₂($j = 0$) reaction at
the collision energy
0.52 kcal/mol. The different
circles represent different HF
product ro-vibrational states.
From [24], reprinted with
permission from AAAS

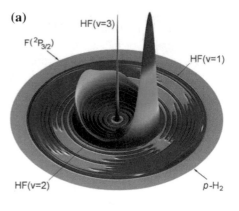

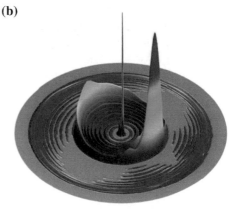

potential energy surface reproduces this peak. Theoretical results show that the
reaction probability for the total angular momentum $J = 0$ exhibits two distinctive
peaks, at 0.26 and 0.46 kcal/mol, which correspond to two reaction resonance
states: the ground and the first excited states (Fig. 3.5). Partial wave analysis
shows that about 65 % of the reaction cross section at 0.52 kcal/mol comes from
the excited resonance with contributions from the $J = 0$ to 4 partial waves,
whereas the other 35 % is due to the ground resonance with contributions from
$J = 5$ to 11 (Fig. 3.6). Clearly, the ground state and the excited state play very
important roles in the reaction, but the excited state plays a major role. In Fig. 3.7,
the contribution of partial waves from the ground and excited states to HF($v' = 2$)
product was further analyzed. Forward scattering of HF($v' = 2$) product is not a
simple sum of the ground state and the excited state contribution, but there is the
effect of instructive interference, while in backward scattering there is destructive
interference.

Figure 3.8 shows theoretical calculated 3D scattering wave functions for $J = 0$
at the collision energy of 0.26 kcal/mol (a) and 0.46 kcal/mol (b). The 3D scattering wave function at 0.26 kcal/mol shows the existence of three nodes along the

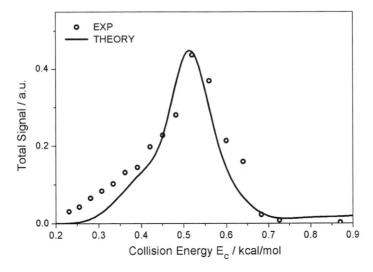

Fig. 3.4 Collision energy-dependent DCS for the forward-scattering $HF(v' = 2)$ product of the reaction $F + H_2(j = 0)$. From [24], reprinted with permission from AAAS

Fig. 3.5 Final vibrational state resolved reaction probabilities for the $F + H_2(j = 0)$ reaction as a function of collision energy for total angular momentum. From [24], reprinted with permission from AAAS

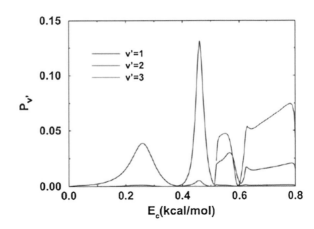

H–F coordinate (correlating to the HF product) in the HF–H′ complex with no node along the reaction coordinate (Fig. 3.8a). This state is the ground state denoted as (003) state. It can be seen from Fig. 3.9 that the (003) state is trapped in the $HF(v' = 3)$-H′ vibrational adiabatic potential (VAP) well. The 3D scattering wave function at 0.46 kcal/mol shows also the existence of three nodes along the H–F coordinate (correlating to the HF product) in the HF–H′ complex with no node along the reaction coordinate (Fig. 3.8b). This state is the excited state denoted as (103) state. The energy of the (103) state is still below the product $HF(v' = 3)$ channel, and it can be seen from Fig. 3.9 that the (103) state is also trapped in the $HF(v' = 3)$-H′ VAP well. This resonance state can be assigned to

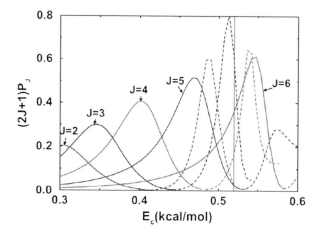

Fig. 3.6 Total reaction probabilities for the F + H₂($j = 0$) reaction as a function of collision energy for total angular momentum $J = 2$–6. For each J *curve*, the *solid* and the *dashed-line* portion represent the contribution from the ground resonance and the excited resonance, respectively. At the *vertical line* (0.52 kcal/mol), only partial waves with $J \geq 5$ of the ground resonance contribute to the reaction, while for the excited resonance, only partial wave from $J = 0$–4 contribute to the reaction. From [24], reprinted with permission from AAAS

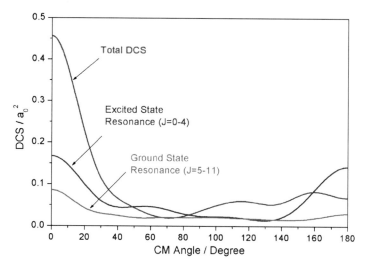

Fig. 3.7 The total DCS of the HF($v' = 2$) product and the DCS contribution to the HF($v' = 2$) product from both the ground and the excited resonance partial waves at 0.52 kcal/mol. From [24], reprinted with permission from AAAS

the (103) resonance state with one-quantum vibration along the reaction coordinate, zero-quantum vibration on the bending motion (or hindered rotation), and three quanta vibration along the HF stretching.

(a)

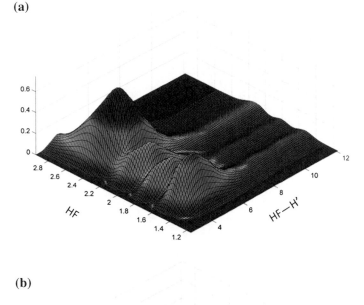

(b)

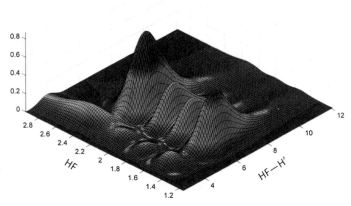

Fig. 3.8 Three dimensional scattering wave functions for $J = 0$ at the collision energy of 0.26 kcal/mol (**a**) and 0.46 kcal/mol (**b**). From [24], reprinted with permission from AAAS

Figure 3.9 shows the resonance-mediated reaction mechanism. The HF($v' = 3$)-H' VAP on the new PES is very peculiar with a deeper vibrational adiabatic well close to the reaction barrier and a shallow van der Waals (vdW) well. The 1D wave function for the ground resonance state in Fig. 3.9 shows that this state is mainly trapped in the inner deeper well of the HF($v' = 3$)-H' VAP with a considerable vdW character, whereas the excited resonance wave function is mainly a vdW resonance. Because of the vdW characters, these two resonance states could likely be accessed via overtone pumping from the HF($v' = 0$)-H' vdW well.

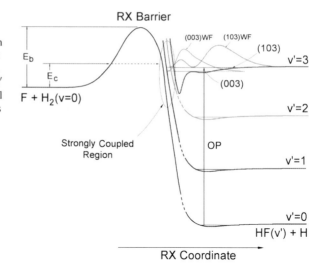

Fig. 3.9 Schematic diagram showing the resonance-mediated reaction mechanism for the F + H₂ reaction with two resonance states trapped in the peculiar HF($v' = 3$)-H' vibrational adiabatic potential well. The 1D wave functions of the two resonance states are also shown. From [24], reprinted with permission from AAAS

Although experimental collision energy is much smaller than the reaction barrier height (2.33 kcal/mol), there is still a big reaction cross section, due to the presence of resonance and its strong quantum tunneling effect. Collision energy at 0.52 kcal/mol is the threshold value of the product HF($v' = 3$) [37]. Due to the experimental collision energy spread, HF($v' = 3$) product will be produced when the collision energy is greater than the threshold value. HF($v' = 2$) and ($v' = 1$) products will be produced when the collision energy is smaller than the threshold value because of strong vibrational nonadiabatic coupling.

3.2.2.2 Effect of H₂ Rotation on Reaction Resonance

We studied the effect of H₂ rotation on reaction resonance, by switching back and forth with normal-H₂ and para-H₂ gas, as previously described.

Figure 3.10 shows the TOF spectra of the H atom product from the F + H₂($j = 0/1$) reaction at the collision energy of 0.56 kcal/mol. Figure 3.11 shows the experimental (a, c) and theoretical (b, d) 3D contour plots for the product HF translational energy and angle distributions for the F + H₂ reaction. The agreement between the theoretical and experimental data is very good. From the results, at the collision energy of 0.56 kcal/mol, F + H₂($j = 0$) product HF($v' = 2$) has a very obvious forward scattering, while F + H₂($j = 1$) product HF($v' = 2$) has almost no forward scattering, but primarily backward and sideways scattering.

Figure 3.12 shows the TOF spectra of the H atom product from the F + H₂($j = 0/1$) reaction at the collision energy of 0.19 kcal/mol. Figure 3.13 shows the experimental (a, c) and theoretical (b, d) 3D contour plots for the product HF translational energy and angle distributions for the F + H₂ reaction.

Fig. 3.10 TOF spectra of the H atom product from the $F + H_2(j = 0/1)$ reaction at the collision energy of 0.56 kcal/mol. TOF spectra at three laboratory angles are shown **a**. $\Theta_L = -60°$ **b**. $\Theta_L = 25°$ **c**. $\Theta_L = 95°$, which correspond roughly to the forward-, sideways-, and backward-scattering directions for the HF($v' = 2$) product in the center-of-mass frame, respectively. Reprinted with permission from [26]. © 2006, American Institute of Physics

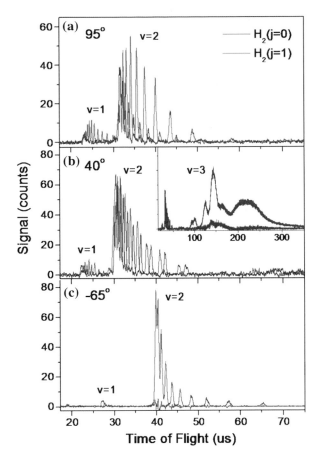

Fig. 3.11 Experimental (**a**, **c**) and theoretical (**b**, **d**) 3D contour plots for the product HF translational energy and angle distributions for the $F + H_2$ reaction at the collision energy 0.56 kcal/mol. The different circles represent different HF product ro-vibrational states. Reprinted with permission from [26]. © 2006, American Institute of Physics

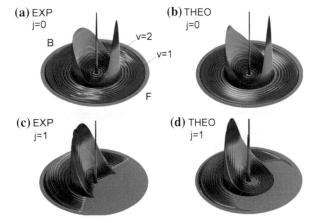

Fig. 3.12 TOF spectra of the H atom product from the F + H₂(j = 0/1) reaction at the collision energy of 0.56 kcal/mol. TOF spectra at three laboratory angles are shown **a.** $\Theta_L = -60°$ **b.** $\Theta_L = 25°$ **c.** $\Theta_L = 95°$, which correspond roughly to the forward-, sideways-, and backward-scattering directions for the HF(v' = 2) product in the center-of-mass frame, respectively. Reprinted with permission from [26]. © 2006, American Institute of Physics

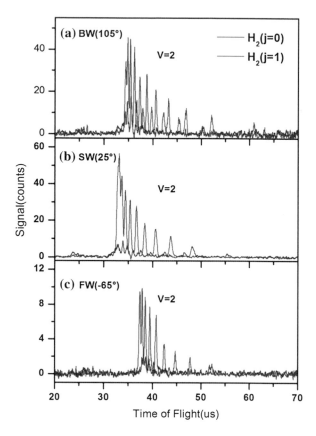

The agreement between the theoretical and experimental data is very good. From the results, at the collision energy of 0.19 kcal/mol, F + H₂(j = 0) product HF(v' = 2) has almost no forward scattering, but primarily backward and sideways scattering, while F + H₂(j = 1) product HF(v' = 2) has a very obvious forward scattering.

The theoretical calculation analyzed the impact of different partial waves on forward and backward scattering. Figure 3.14 shows the angular distribution of the reaction F + H₂(j = 0) product HF(v' = 2) from the contribution of the ground resonance, the excited resonance, and the continuous tunneling partial waves at 0.56 kcal/mol. The continuum tunneling reaction mechanism plays a more important role at the collision energy of 0.56 kcal/mol, mainly representing as backward scattering. In forward scattering, there is a strong destructive interference between the ground and excited resonances at this collision energy. Compared with the reaction at the collision energy of 0.52 kcal/mol, although the collision energy is changed only 0.04 kcal/mol ($\sim$14 cm^{-1}), the forward-scattering intensity is greatly changed. There is a dramatic relative phase shift between the two resonance pathways at the two collision energies. But in F + H₂(j = 1) at

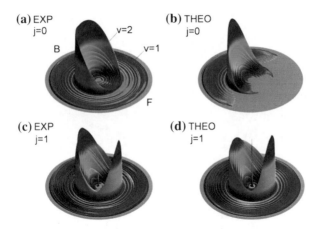

Fig. 3.13 Experimental (**a, c**) and theoretical (**b, d**) 3D contour plots for the product HF translational energy and angle distributions for the F + H$_2$ reaction at the collision energy 0.19 kcal/mol. The different circles represent different HF product ro-vibrational states. Reprinted with permission from [26]. © 2006, American Institute of Physics

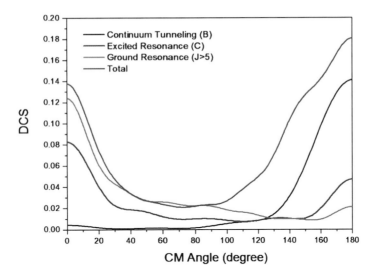

Fig. 3.14 The angular distribution of the HF($v' = 2$) product from the contribution of the ground resonance, the excited resonance, and the continuous tunneling partial waves at 0.56 kcal/mol

the collision energy of 0.19 kcal/mol, there is a strong constructive interferences between the partial waves of the different total angular momentum, as shown in Fig. 3.15.

Another dynamical effect of H$_2$ rotation is at two collision energies, HF($v' = 2$) product corresponding to H$_2$($j = 1$) has a high rotational excitation, while

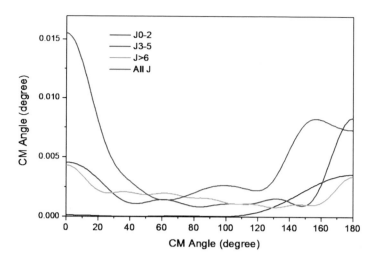

Fig. 3.15 The angular distribution of the HF($v' = 2$) product from the contribution of the ground resonance, the excited resonance, and the continuous tunneling partial waves at 0.56 kcal/mol. Reprinted with permission from [26]. © 2006, American Institute of Physics

HF($v' = 2$) product corresponding to $H_2(j = 0)$ has a much smaller rotational excitation. Theoretical analysis of the wavepacket flux at the entrance channel shows that the $F + H_2(j = 1)$ reaction access the bent TS more efficiently than the $H_2(j = 0)$ reaction [26]. A correlation is also found between the averaged incident angle of the wavepacket flux and the HF product rotational excitation. This seems to indicate that the reagent rotationally excitation effect on the HF product rotational distribution is related to the stereodynamics of this reaction system.

Effects of H_2 rotation on $F + H_2$ reaction are completely different at different collision energies of 0.56 kcal/mol and 0.19 kcal/mol. At 0.56 kcal/mol, the $F + H_2(j = 0)$ reaction has an obvious forward scattering, while the $F + H_2(j = 1)$ reaction has primarily backward scattering; at 0.19 kcal/mol, the $F + H_2(j = 0)$ reaction has primarily backward scattering, while the $F + H_2(j = 1)$ reaction has an obvious forward scattering. Because $H_2(j = 1)$ molecules provide 118 cm^{-1} (~ 0.34 kcal/mol) energy, the $F + H_2(j = 1)$ reaction exceeds the reaction resonance region at the collision energy of 0.56 kcal/mol. Therefore, the product exhibits a significantly backward scattering. At 0.19 kcal/mol, reaction resonance just occurs in the $F + H_2(j = 1)$ reaction, resulting in a very obvious forward scattering; the $H_2(j = 0)$ reaction is below the first resonance state, so it has primarily backward scattering, due to the contribution of continuum tunneling reaction mechanism. Figure 3.16 shows a schematic diagram of the reaction mechanism.

In order to look for more strong experimental evidences in the $F + H_2(j = 1)$ reaction resonances, we have measured the collision energy dependent forward-scattering signal (see Fig. 3.17) of the HF($v' = 2$) product from the $F + H_2(j = 1)$ reaction in the collision energy range of 0.13–0.31 kcal/mol. The main peak of the distribution is located at the collision energy of ~ 0.18 kcal/mol. Full quantum

Fig. 3.16 Schematic diagram showing the resonance-mediated reaction mechanism for the $F + H_2$ reaction with two resonance states trapped in the peculiar $HF(v' = 3)$-H' vibrational adiabatic potential well. The 1D wave functions of the two resonance states are also shown

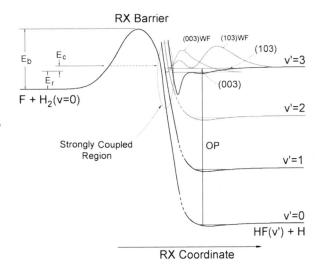

theoretical scattering result, shown in Fig. 3.17, is in rather good agreement with the experimental. Further theoretical analysis attributed this main peak at ~ 0.18 kcal/mol to the two Feshbach resonances of the $F + H_2(j = 1)$ reaction. In addition to the main peak, a shoulder in the high energy side (~ 0.28) seems to be present, this can be assigned to the $J = 6-7$ partial wave scattering of the ground Feshbach resonance. Experimental result also seems to be in support of the existence of this shoulder. This study clearly shows how the reactant rotational

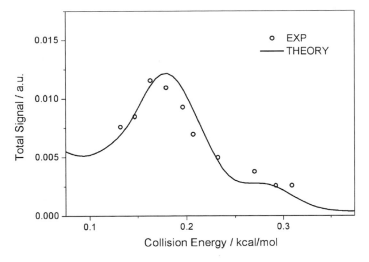

Fig. 3.17 Collision energy-dependent DCS for the forward-scattering $HF(v' = 2)$ product of the reaction $F + H_2(j = 1)$. Reprinted with permission from [26]. © 2006, American Institute of Physics

state affects reaction resonance; in addition, it once again interprets the reaction resonance.

We also carried out research on F + H₂ → HF + H reaction at other collision energies, see Ref. [23].

3.2.3 Summary

We systematically studied the $F(^2P_{3/2})$ + H₂ → HF + H reaction and found that the product HF($v' = 2$) has obvious forward scattering. The experimental finding of reaction resonance in this system, combined with the theoretical calculations from Daiqian Xie and Donghui Zhang et al. gave a perfect answer to the question whether reaction resonance exists in this system. We also pictured resonance reaction mechanism for this system. Moreover, we studied the dynamical effects of the H₂ rotation in the reaction, analyzed how the H₂ rotation affects reaction resonance, and gave a clear physical picture, to further enrich the content of the reaction resonance.

3.3 Reaction Resonance Phenomenon in $F(^2P_{3/2})$ + HD → HF + D

Isotope substitution is a very effective way to test the precision and accuracy of the potential energy surface. Kopin Liu and Skodje have studied in detail the F + HD → HF + D reaction and proved the existence of reaction resonance phenomenon in this system, experimentally and theoretically [7, 10, 29, 30]. In order to obtain more accurate experimental data, we measured this reaction precisely once again and cooperated with theoretical coworkers Donghui Zhang et al. We obtained potential energy surface with the spectral accuracy, also observed some new phenomena never found before.

3.3.1 Experimental Method

We used a method similar to in the F + H₂ reaction to study the F + HD → HF + D reaction with the collision energy changing from 0.3 to 1.2 kcal/mol. In the experiment, a fluorine atom beam is generated by a double-stage pulsed discharge [27]. In the experiments, there were no obvious products from F* reaction. HD molecular beam is generated by HD gas (backpressure ∼2 atm) through supersonic expansion with a general valve cooled by liquid nitrogen (∼78 K). After supersonic cooling, rotational temperature of the beam is very low, almost all of the HD are populated at the lowest rotational HD($v = 0, j = 0$) state. In the experiment, there was no obvious signal from HD($v = 0, j = 1$) reaction. For experiments at low collision energies, we mixed HD with the heavier Ne to reduce the HD molecular beam speed in order to obtain low collision energies. We changed the collision energy by changing the collision angle. TOF spectra were measured at many laboratory angles at $\approx 10°$ intervals in the range from 0° to 180° in the center-of-mass frame at each collision energy. The TOF spectra were fitted to ro-vibrational resolved differential cross section in the center-of-mass frame. The reaction energy for F + HD($j = 0$) → HF + D reaction is 10,905.0 cm^{-1}.

3.3.2 Experimental Results

3.3.2.1 Time-of-Flight Spectra of the D Atom Products

Time-of-flight (TOF) spectra of the D atom products from the F + HD reaction were measured at many laboratory angles at $\approx 10°$ intervals, with the collision energy changing from 0.3 to 1.2 kcal/mol. Here, TOF spectra of the D atom products at three (or two) laboratory angles at 11 collision energies are shown in from Figs. 3.17, 3.18, 3.19, 3.20, 3.21, 3.22, 3.23, 3.24, 3.25, 3.26, 3.27, and 3.28, corresponding roughly to the backward, sideways, and forward-scattering directions for the HF($v' = 2$) product in the center-of-mass frame, respectively. In the TOF spectra, a clear bimodal structure in the HF($v' = 2$) rotational distribution is shown. It will be discussed later in detail.

Fig. 3.18 TOF spectra of the D atom product from the F + HD($j = 0$) reaction at the collision energy of 0.31 kcal/mol

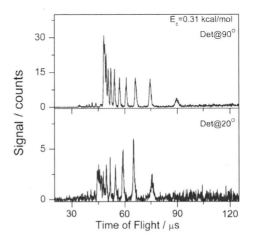

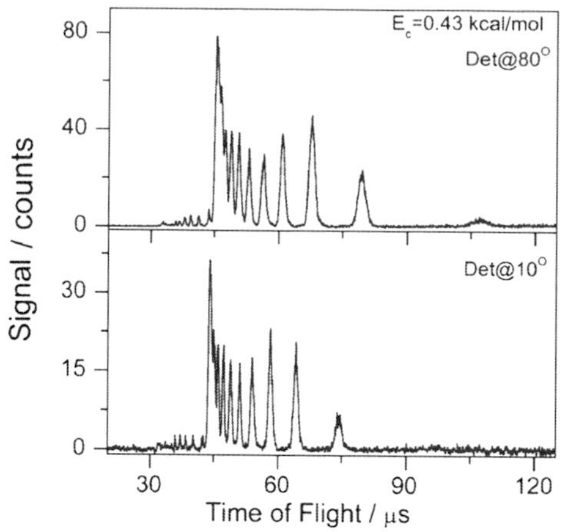

Fig. 3.19 TOF spectra of the D atom product from the F + HD($j = 0$) reaction at the collision energy of 0.43 kcal/mol. Reprinted from [34]. © 2008, National Academy of Sciences, USA

Fig. 3.20 TOF spectra of the D atom product from the F + HD($j = 0$) reaction at the collision energy of 0.48 kcal/mol. Reprinted from [34]. © 2008, National Academy of Sciences, USA

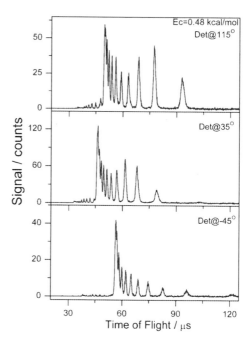

3.3.2.2 Differential Cross Section

The D atom TOF spectra were fitted to DCS in the center-of-mass frame. DCS at 11 collision energies are shown below. The most intriguing observation in the experimental DCS is the dramatic change of the 3D DCS. At collision energy of 0.43 kcal/mol, the products have mainly backward scattering; at 0.48 kcal/mol, the products move to sideways, backward scattering is reduced and small amount of forward scattering appears; at 0.52 kcal/mol, forward scattering continues

Fig. 3.21 TOF spectra of the D atom product from the F + HD($j = 0$) reaction at the collision energy of 0.52 kcal/mol. Reprinted from [34]. © 2008, National Academy of Sciences, USA

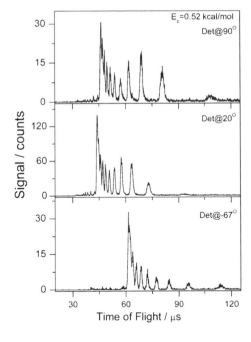

Fig. 3.22 TOF spectra of the D atom product from the F + HD($j = 0$) reaction at the collision energy of 0.59 kcal/mol

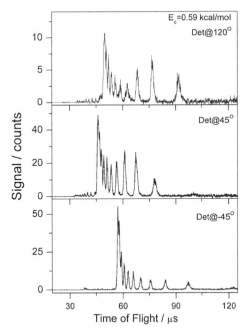

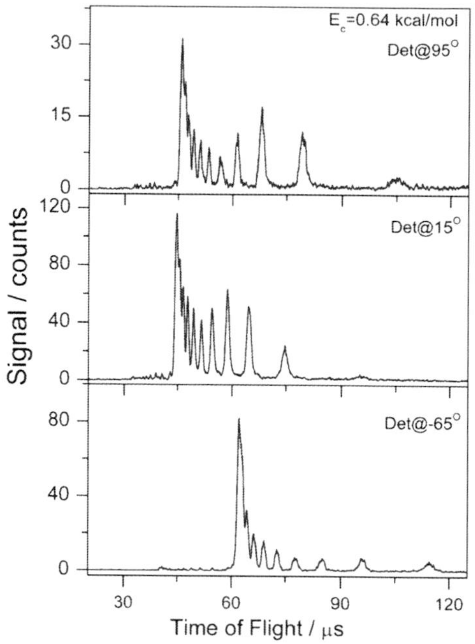

Fig. 3.23 TOF spectra of the
D atom product from the
F + HD(j = 0) reaction at
the collision energy of
0.64 kcal/mol

Fig. 3.24 TOF spectra of the
D atom product from the
F + HD(j = 0) reaction at
the collision energy of
0.71 kcal/mol. Reprinted
from [34]. © 2008, National
Academy of Sciences, USA

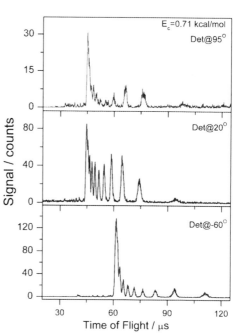

Fig. 3.25 TOF spectra of the
D atom product from the
F + HD($j = 0$) reaction at
the collision energy of
0.85 kcal/mol

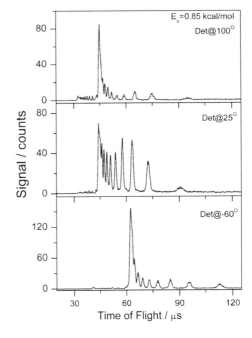

Fig. 3.26 TOF spectra of the
D atom product from the
F + HD($j = 0$) reaction at
the collision energy of
0.95 kcal/mol

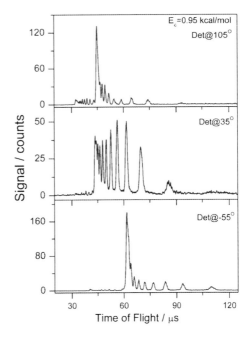

Fig. 3.27 TOF spectra of the
D atom product from the
F + HD($j = 0$) reaction at
the collision energy of
0.85 kcal/mol

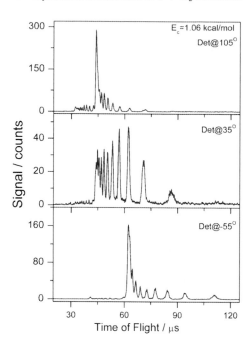

Fig. 3.28 TOF spectra of the
D atom product from the
F + HD($j = 0$) reaction at
the collision energy of
1.17 kcal/mol

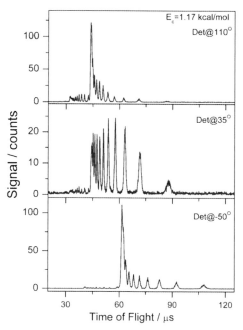

becoming larger, and the products have mainly sideways-backward and forward scattering; at 0.59 and 0.64 kcal/mol, the products have significant enhancement in forward scattering; at 0.71 kcal/mol, the products have mainly forward scattering, with rare sideways and backward scattering. There are obvious differences between the results of the current experiment and those from Kopin Liu and coworkers at the collision energy 0.5 kcal/mol. The possible reason is that Liu et al. had a certain deviation on the F atom speed determination. Here, the calculated theoretical DCS results at 11 collision energies are also shown, which are in good agreement with our experimental results.

3.3.2.3 Collision Energy Dependence of the Backward-Scattering Signal

To further investigate the resonance phenomenon in the F + HD($j = 0$) → HF + D reaction, we have also measured the scattering signal at the exact backward-scattering direction for HF($v' = 2$) product in low rotational states ($j' = 0$–3) in the collision energy range between 0.2 and 1.2 kcal/mol. Figure 3.29 shows the summed signal of HF($v' = 2$, $j' = 0$–3) as a function of collision energy. The experimental results show a clear peak ∼0.39 kcal/mol for the backward-scattered HF($v' = 2$, $j' = 0$–3) products. The peak in the collision energy-dependent backward-scattered HF($v' = 2$) product is obviously related to the resonance phenomenon in the reaction [25]. Figure 3.29 shows also calculated

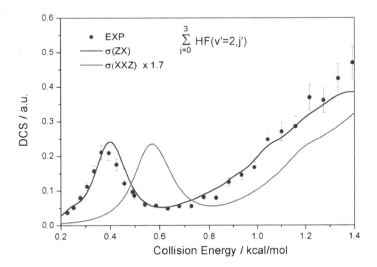

Fig. 3.29 Collision energy-dependent DCS for the backward-scattering HF($v' = 2$) products summed over at $j' = 0$–3. The *solid circles* are the experimental data, and the *solid lines* are the calculated theoretical results based on the XXZ-PES and the new CCSD(T)-PES. Reprinted from [34]. © 2008, National Academy of Sciences, USA

results on the XXZ potential energy surface and the new CCSD(T) potential
energy surface from Donghui Zhang et al. Calculated DCS on the XXZ-PES is in
strong disagreement with the present experimental result, with a ~0.2 kcal/mol
shift. However, calculated DCS on the CCSD(T)-PES is in strong agreement with
the experimental result. It will be discussed later in detail.

3.3.2.4 HF(v' = 2) Product Rotational Population

From the TOF spectra, the rotational distribution of HF(v' = 2) appears dimodal.
To better understand the rotational distribution of the products, we multiply the
DCS of each rotational state by $\sin\theta$ (θ is the product scattering angle in the center-
of-mass coordinate system) and integrate θ ranging from 0° to 180° (we detected
only the section along the reactant relative velocity direction in the whole scattering velocity space), thus get relative integral cross sections. Plotting these relative integral cross sections to HF(v' = 2) quantum number j', the HF(v' = 2)
product rotational state distribution can be directly seen, as shown in Fig. 3.30.
From the present experiment, we have found that the rotational distribution of the

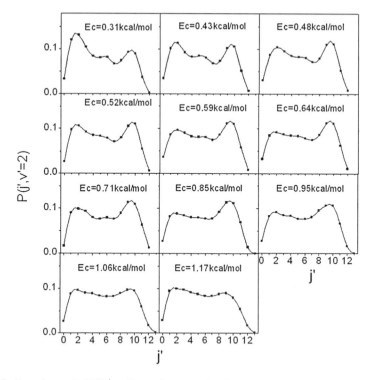

Fig. 3.30 Experimental HF(v' = 2) product rotational state distribution from the reaction
F + HD at a series of collision energies

HF($v' = 2$) product from the F + HD reaction is also very intriguing. At the lowest collision energies studied from 0.31 to 0.64 kcal/mol, the distribution appears trimodal with two clear peaks at $j' = 2$ and $j' = 9$, with another small bump around $j' = 6$ (see Fig. 3.33). At higher collision energies, the small middle bump becomes less obvious, while the two main peaks persist. The distribution observed in this work is quantitatively different from the previous experimental result at similar collision energies for the same system by Kopin Liu et al. mainly due to the much higher resolution and accuracy of our experiments. Clearly, the rotational distribution observed in F + HD is considerably different from that of the F + H$_2$ reaction. The dynamical origin of this trimodal distribution is not immediately clear at this moment and certainly deserves more detailed investigations [25].

3.3.3 Discussion

As previously described above, XXZ-PES constructed by Donghui Zhang and coworkers is a good explanation to the reaction resonance phenomenon in the F + H$_2$ reaction. However, as shown in Fig. 3.29, calculations, based on the XXZ-PES, could not model the experimental results of collision energy dependence of product D atom backward-scattering signal in the F + HD($j = 0$) → HF + D reaction. There was a large deviation from the experimental data. Clearly, XXZ-PES is still not accurate enough.'

Donghui Zhang and coworkers constructed a new CCSD(T)-PES and carried out full quantum dynamical calculations based on the CCSD(T)-PES for the F + HD($j = 0$) → HF + D reaction, see reaction DCS in Fig. 3.31. For the F + HD reaction, the collision energy dependence for the backward-scattering HF($v' = 2, j' = 0$–3) product calculated on the new PES is shifted down to lower energy by ∼0.16 kcal/mol compared with that on the XXZ-PES, giving rise to a nearly perfect agreement with the experimental result as shown in Fig. 3.29. The experimental DCS varies dramatically in a very small collision energy range, as shown in Fig. 3.32. Remarkably, the theoretical results obtained on the new PES can reproduce closely this dramatic variation in DCS as shown in Fig. 3.31. Furthermore, the theory on the new PES is able to reproduce almost all of the finest structures in the DCSs observed experimentally, in particular the intriguing trimodal structures in the HF($v' = 2, j'$) rotational distribution at the low collision energies (see Fig. 3.33). But the dynamical origin of this trimodal distribution is still unclear [25].

The new PES can be a good description for the F + HD, but is it accurate for the F + H$_2$ reaction? The collision energy dependence for the forward-scattering HF($v' = 2$) product obtained on the new PES agrees with the experimental results very well, with a clear narrow peak predicted at 0.52 kcal/mol (see Fig. 3.34). The degree of the agreement on the DCSs between experiment and theory is also remarkable.

Fig. 3.31 The calculation results of three-dimensional D atom product flux contour plots base on CCSD(T)-PES for the
F + HD($j = 0$) → HF(v', j') + D reaction. Reprinted from [34]. © 2008, National Academy of Sciences, USA

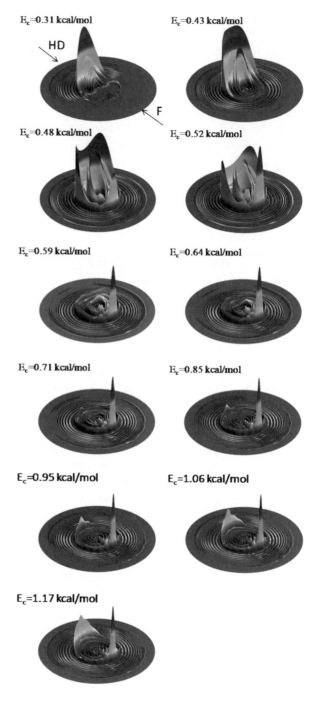

Fig. 3.32 The experimental results of three-dimensional D atom product flux contour plots for the $F + HD(j = 0) \rightarrow HF(v', j') + D$ reaction. Reprinted from [34]. © 2008, National Academy of Sciences, USA

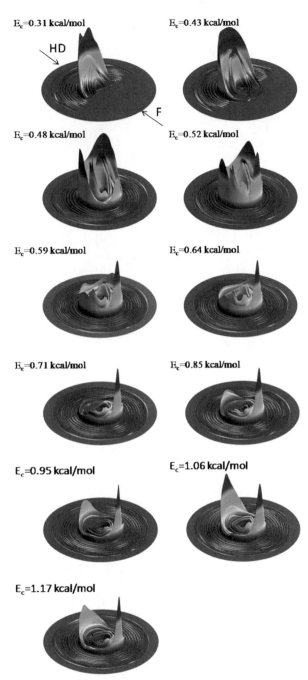

Fig. 3.33 The HF($v' = 2$) product rotational distribution from the reaction of F + HD at the collision energy of 0.48 kcal/mol. Experimental data (*red*) and theoretical results from full quantum dynamical calculations based on the CCSD(T)-PES are shown. Reprinted from [34]. © 2008, National Academy of Sciences, USA

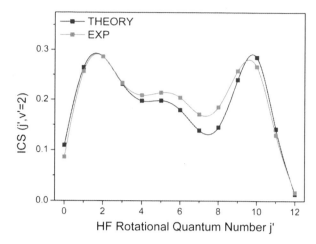

Fig. 3.34 Collision energy-dependent DCS for the forward-scattering HF($v' = 2$) product of the reaction F + H₂($j = 0$) and the theoretical results based on the CCSD(T)-PES

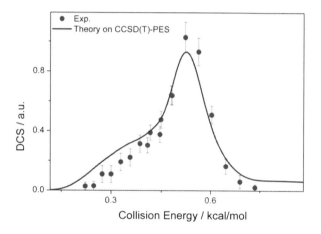

Further theoretical analysis reveals that all these fascinating phenomena observed at low collision energy for the F + HD → HF + D reaction are directly related to the ground resonance state trapped in the peculiar HF($v' = 3$)-D adiabatic well, denoted as (003) state, as shown in Fig. 3.35b in one dimension along the reaction coordinate. Interestingly, for the total angular momentum $J = 0$, the energy for that resonance state, 0.40 kcal/mol, is rather close to the peak position (~ 0.39 kcal/mol) of the backward-scattering HF($v' = 2, j' = 0$–3) signal, because the backward-scattering product is mainly produced by the low total angular momentum partial waves. As shown in Fig. 3.35b, the biggest difference between the new PES and the XXZ-PES is the depth of the potential well. The HF($v' = 3$)-D adiabatic potential on the new PES is almost identical to that on the XXZ-PES when the distance between HF and D is larger than 5.5 bohr. However, for smaller values of the distance, the adiabatic potential on the new PES is considerably different from that on the XXZ-PES. The well of the new PES is deeper than that

Fig. 3.35 a The one-dimensional adiabatic resonance potentials of HF($v = 3$)–H for the F + H$_2$ reaction traced out from the XXZ-PES and the new PES. **b** The one-dimensional adiabatic resonance potentials of HF($v' = 3$)–D for the F + HD reaction traced out from the XXZ-PES and the new PES. Reprinted from [34]. © 2008, National Academy of Sciences, USA

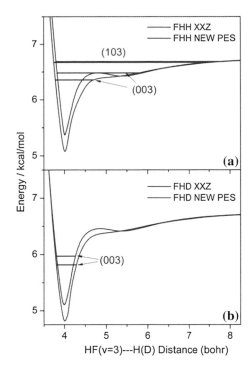

on the XXZ-PES by ∼0.3 kcal/mol. And the small bump between the inner well and the vdW well at the distance of 4.5 bohr on the XXZ-PES vanishes on the new PES. However, a shoulder in the vdW interaction region is present on the new potential, suggesting that the vdW interaction is still important in this system. Consequently, the energy of the ground resonance state on the new PES shifts down by ∼0.16 kcal/mol, reaching 0.4 kcal/mol [25].

The implication of the new PES on the resonance states for F + H$_2$ reaction is shown in Fig. 3.35a. The energy shift for the ground resonance state (003) of 0.12 kcal/mol is quite substantial. However, because the energy of the (103) state is mainly determined by the potential at large distance, this state shifts down by merely 0.01 kcal/mol. The phenomenon of predominantly forward-scattering HF($v' = 2$) product at the collision energy of 0.52 kcal/mol is a result of the interference between the (003) and (103) states [24], and it is largely determined by the (103) state. Because both the new PES and the XXZ-PES are at the same level of accuracy in describing the (103) state, both of them give the correct forward-scattering HF($v' = 2$) product signal [25].

Clearly, the new PES not only reproduces the experimental results for the F + H$_2$ → HF + H reaction, but also agrees with the experimental results very well, proving its accuracy in the TS region. The combination of experiments and theoretical calculation can deepen our understanding of the nature of the chemical reaction.

3.3.4 Summary

Using high-resolution crossed molecular beam, we studied the $F(^2P_{3/2}) + HD \rightarrow$ HF + D reaction. We carried out accurate measurements of the differential cross section at 11 collision energies ranging from 0.3 to 1.2 kcal/mol and observed a dramatic change of the 3D DCS at collision energy around 0.5 kcal/mol. We also found an obvious peak of product backward-scattering signal in the collision energy dependence and multi-peak structure of the product HF($v' = 2$) rotational state distribution. The experimental results proved the accuracy of the new CCSD(T)-PES by Donghui Zhang and coworkers with spectroscopic accuracy (a few cm^{-1}). Through the combination of detailed research of isotopic substitution reaction and theoretical calculation, we provided an accurate PES for this benchmark reaction.

3.4 Conclusion

Through high-resolution, highly accurate crossed beam scattering experiments, we studied the $F(^2P_{3/2}) + H_2(j = 0) \rightarrow$ HF + H reaction. Cooperating with colleagues working on the theoretical calculation, for the first time, we demonstrated clearly the presence of reaction resonance phenomenon, ending the controversy whether the resonance exists in this reaction, which puzzled the academia for more than 30 years. We also studied the effects of reactant H$_2$ rotation on reaction resonance, explained the mechanism of the reaction resonance, and gave a clear physical picture.

Using isotope substitution, we studied the $F(^2P_{3/2}) + HD \rightarrow$ HF + D reaction and carried out accurate measurement of the differential scattering cross section at 11 collision energies ranging from 0.3 to 1.2 kcal/mol, and collision energy dependence of product backward-scattering signal. We provided highly accurate experimental results for theoretical calculation and proved the accuracy of the new CCSD(T)-PES with spectroscopic accuracy. This is the most accurate PES for this system so far.

References

1. Althorpe SC, Clary DC (2003) Quantum scattering calculations on chemical reactions. Annu Rev Phys Chem 54:493–529
2. Aoiz FJ, Banares L, Herrero VJ (1998) Recent results from quasiclassical trajectory computations of elementary chemical reactions. J Chem Soc-Faraday Trans 94:2483–2500
3. Aoiz FJ, Banares L, Herrero VJ et al (1994) Classical dynamics for the F + H$_2$ → HF + H reaction on a new ab initio potential energy surface. A direct comparison with experiment. Chem Phys Lett 223:215–226

4. Aquilanti V, Cavalli S, De Fazio D et al (2002) Exact reaction dynamics by the hyperquantization algorithm: integral and differential cross sections for F + H₂, including long-range and spin orbit effects. Phys Chem Chem Phys 4:401–415

5. Castillo JF, Manolopoulos DE, Stark K et al (1996) Quantum mechanical angular distributions for the F + H₂ reaction. J Chem Phys 104:6531–6546

6. Che L (2008) thesis: studies on non-adiabatic effect and reaction resonance in the reaction F(^{2}P) + H₂/HD/D₂. Institute of Chemical Physics, CAS. Dalian China

7. Dong F, Lee SH, Liu K (2000) Reactive excitation functions for F + p-H₂/n-H₂/D₂ and the vibrational branching for F + HD. J Chem Phys 113:3633–3640

8. Hayes M, Gustafsson M, Mebel AM et al (2005) An improved potential energy surface for the F + H₂ reaction. Chem Phys 308:259–266

9. Hu QJ, Hepburn JW (2006) Energetics and dynamics of threshold photoion-pair formation in HF/DF. J Chem Phys 124:74311

10. Lee SH, Dong F, Liu KP (2006) A crossed-beam study of the F + HD → HF + D reaction: the resonance-mediated channel. J Chem Phys 125:133106

11. Lee SH, Dong F, Liu KP (2002) Reaction dynamics of F + HD → HF + D at low energies: resonant tunneling mechanism. J Chem Phys 116:7839–7848

12. Lee YT (1986) Molecular beam studies of elementary chemical processes. Nobel Lecture

13. Lee YT (1987) Molecular beam studies of elementary chemical processes. Science 236(4803):793–798

14. Mann DE, Thrush BA, Lide JDR et al (1961) Spectroscopy of fluorine flames. I. Hydrogen-fluorine flame and the vibration-rotation emission spectrum of HF. J Chem Phys 34:420–431

15. Manolopoulos DE (1997) The dynamics of the F + H₂ reaction. J Chem Soc-Faraday Trans 93:673–683

16. Manolopoulos DE, Stark K, Werner HJ et al (1993) The transition-state of the F + H₂ reaction. Science 262:1852–1855

17. Neumark DM, Wodtke AM, Robinson GN et al (1984) Experimental investigation of resonances in reactive scattering: The F + H₂ reaction. Phys Rev Lett 53:226

18. Neumark DM, Wodtke AM, Robinson GN et al (1985) Molecular beam studies of the F + H₂ reaction. J Chem Phys 82:3045–3066

19. Neumark DM, Wodtke AM, Robinson GN et al (1985) Molecular beam studies of the F + D₂ and F + HD reactions. J Chem Phys 82:3067–3077

20. Parker JH, Pimentel GC (1969) Vibrational energy distribution through chemical laser studies. I. Fluorine atoms plus hydrogen or methane. J Chem Phys 51:91–96

21. Polanyi JC, Tardy DC (1969) Energy distribution in the exothermic reaction F + H₂ and the endothermic reaction HF + H. J Chem Phys 51:5717–5719

22. Polanyi JC, Woodall KB (1972) Energy distribution among reaction products. VI. F + H₂, D₂. J Chem Phys 57:1574–1586

23. Qiu M (2006) thesis: High resolution crossed molecular beams study on the F + H₂ reaction. Dalian Institute of Chemical Physics, CAS. Dalian, China

24. Qiu M, Ren Z, Che L et al (2006) Observation of Feshbach resonances in the F + H₂ → HF + H reaction. Science 311:1440–1443

25. Ren Z, Che L, Qiu M et al (2008) Probing the resonance potential in the F atom reaction with hydrogen deuteride with spectroscopic accuracy. Proc Natl Acad Sci USA 105:12662–12666

26. Ren ZF, Che L, Qiu MH et al (2006) Probing Feshbach resonances in F + H-2(j = 1) → HF + H: Dynamical effect of single quantum H₂-rotation. J Chem Phys 125

27. Ren ZF, Qiu MH, Che L et al (2006) A double-stage pulsed discharge fluorine atom beam source. Rev Sci Instrum 77:016102

28. Schatz GC, Bowman JM, Kuppermann A (1973) Large quantum effects in the collinear F + H₂ → FH + H reaction. J Chem Phys 58:4023–4025

29. Skodje RT, Skouteris D, Manolopoulos DE et al (2000) Resonance-mediated chemical reaction: F + HD → HF + D. Phys Rev Lett 85:1206

30. Skodje RT, Skouteris D, Manolopoulos DE et al (2000) Observation of a transition state resonance in the integral cross section of the F + HD reaction. J Chem Phys 112:4536–4552

31. Stark K, Werner HJ (1996) An accurate multireference configuration interaction calculation of the potential energy surface for the F + H$_2$ → HF + H reaction. J Chem Phys 104:6515–6530
32. Stoicheff BP (2001) On the dissociation energy of molecular hydrogen. Can J Phys 79:165–172
33. Takayanagi T (2006) The effect of van der Waals resonances on reactive cross sections for the F + HD reaction. Chem Phys Lett 433:15–18
34. Wang XG, Dong WR, Qiu MH et al (2008) HF(v' = 3) forward scattering in the F + H$_2$ reaction: Shape resonance and slow-down mechanism. Proc Natl Acad Sci USA 105:6227–6231
35. Wu S-F, Johnson BR, Levine RD (1973) Quantum mechanical computational studies of chemical reactions: III. Collinear A BC reaction with some model potential energy surfaces. Mol Phys 25:839–856
36. Xu CX, Xie DQ, Zhang DH (2006) A global ab initio potential energy surface for F + H$_2$ → HF + H. Chin J Chem Phys 19:96–98
37. Yang XF, Hwang DW, Lin JJ et al (2000) Dissociation dynamics of the water molecule on the (A)over-tilde B-1(1) electronic surface. J Chem Phys 113:10597–10604

Chapter 4
The Non-adiabatic Effects
in $F(^2P) + D_2 \rightarrow DF + D$

Using HRTOF crossed molecular beam technique, we studied the $F(^2P) + D_2 \rightarrow$ DF + D reaction systematically. The differential cross sections and the relative integral cross sections in this reaction were provided. We measured the $F(^2P_{3/2})/$ $F*(^2P_{1/2})$ ratio in the F atom beam and thus got the $F(^2P_{3/2})/F*(^2P_{1/2})$ reactivity ratio, which is in perfect agreement with the theoretical results. In Sect. 4.1 of this chapter, we review non-adiabatic effects in the $F + H_2$ system; the measurement of the $F(^2P_{3/2})/F*(^2P_{1/2})$ ratio in the F atom beam is described in Sects. 4.2, 4.3 presents the crossed molecular beam experimental results and discussion; summary is given in the last section.

4.1 Review of Non-adiabatic Effects in the $F + H_2$ System

Figure 4.1 shows a schematic of relative energy in the $F(^2P) + H_2(j = 0)$ reaction. Due to the spin–orbit interaction, the fluorine atom degenerate ground electronic state $F(^2P)$ is split into two states: the spin–orbit ground state $F(^2P_{3/2})$ and the spin–orbit excited state $F*(^2P_{1/2})$, respectively. As shown in Fig. 4.1, on the three adiabatic potential energy surfaces, electronic states $1^2A'$ and $1^2A''$ ($^2\Sigma^+_{1/2}$ and $^2\Pi_{3/2}$ in linear configuration) are adiabatically related to the ground state $F(^2P_{3/2})$, electronic state $2^2A'$ ($^2\Pi_{1/2}$ in linear configuration) is adiabatically related to the excited state $F*(^2P_{1/2})$. Among them, only electronic state $2^2A'$ is adiabatically related to the product ground electronic state $[HF(X^1\Sigma^+) + H(^2S)]$, the other two electronic states are adiabatically related to excited electronic state $[HF(a^3\Pi) + H(^2S)]$. However, the energy of excited state $HF(a^3\Pi)$ is very high, and it is impossible to achieve the collision energy under the general experimental conditions. Within the B–O approximation, the nuclei moves only on a single potential energy surface, so only the fluorine atom spin–orbit ground state $F(^2P_{3/2})$ can react, while reaction of excited state $F*(^2P_{1/2})$ is forbidden. The potential energy surface of $F(^2P) + HD$ and $F(^2P) + D_2$ is similar to that of $F(^2P) + H_2$, while the excited state $F*(^2P_{1/2})$ is forbidden within the B–O approximation.

Z. Ren, *State-to-State Dynamical Research in the F+H₂*
Reaction System, Springer Theses, DOI: 10.1007/978-3-642-39756-1_4,
© Springer-Verlag Berlin Heidelberg 2014

Fig. 4.1 Schematic of relative energy in the reaction F(^{2}P) + H$_2$

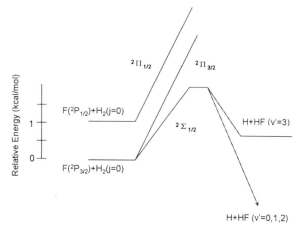

However, due to the non-adiabatic effects, coupling occurs between potential energy surfaces, B–O approximation breaks down, allowing the reaction of F*(^{2}P$_{1/2}$). Interest in the reactivity of spin–orbit excited F atom with H$_2$, induced by the breakdown of the B–O approximation, dates back 30 years [1]. Earlier experimental studies of the F + H$_2$ [2] and F + D$_2$ [3, 4] reactions showed no concrete evidence of F*(^{2}P$_{1/2}$) reactivity. The great experimental difficulty in evaluating the relative F*(^{2}P$_{1/2}$) reactivity is the separation of HF products produced by the reaction of F*. Nesbitt and coworkers [5–7] did find such evidence by looking at the production of HF($v' = 3$) in rotational levels that are energetically inaccessible by reaction of ground spin–orbit state F atoms. Concurrently, in a crossed beams study of the F/F* + HD reaction, Liu and coworkers also found experimental evidence of the HF($v' = 3$) product from the F*(^{2}P$_{1/2}$) reaction at threshold collision energies [8]. However, they did not give the differential cross section of the F*(^{2}P$_{1/2}$) reaction. Because the F(^{2}P$_{3/2}$)/F*(^{2}P$_{1/2}$) ratio in the F atom beam source is unknown, the F*(^{2}P$_{1/2}$)/F(^{2}P$_{3/2}$) reactivity ratio cannot be quantitatively measured.

Despite the importance of F*(^{2}P$_{1/2}$) reactivity, the majority of prior theoretical work on the F + H$_2$ (HD, D$_2$) reaction was limited to calculations on the lowest, electronically adiabatic PES. Building on earlier work, for the first time Alexander and Werner [9, 10] presented an exact time-independent framework for the fully quantum mechanical study of this reaction, as well as spin–orbit and Coriolis coupling. Recently, Han and coworkers [11–13] studied the F(^{2}P) + H$_2$ system using time-dependent quantum wavepacket method. They all have shown that, as the collision energy decreases, F*(^{2}P$_{1/2}$) reaction becomes increasingly important because of the additional internal energy available in the spin–orbit excited state. At low energy, the reaction cross section of F*(^{2}P$_{1/2}$) is greater than that of F(^{2}P$_{3/2}$).

In this chapter, we carried out detailed crossed molecular beam study of the F(^{2}P) + o−D$_2$(j = 0) → DF + D reaction. Figure 4.2 shows schematic of relative energy in the reaction F(^{2}P) + D$_2$. The choice of D$_2$, rather than H$_2$ or HD, was

Fig. 4.2 Schematic of relative energy in the reaction $F(^2P) + D_2$. From [25], reprinted with permission from AAAS

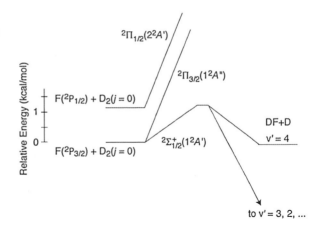

made for several reasons: The absence of D atom background allows a higher detection signal-to-noise ratio, and the smaller rotational constant of DF (compared with HF) allows a cleaner separation of the DF (v') products from the reaction of F compared with F*. In addition, because of the absence of a transition state resonance in the F + D_2 reaction at low collision energy, the investigation and the comparison with theory can focus solely on the relative reactivity of the two F atom spin–orbit states, without any additional complexity introduced by the resonance.

4.2 Measurement of the $F(^2P_{3/2})$ and $F*(^2P_{1/2})$ Ratio in the Fluorine Atom Beam

In recent years, resonance-enhanced multiphoton ionization (REMPI) [14, 15] and Stern–Gerlach magnetic analysis techniques [16, 17] are used to measure the $F(^2P_{3/2})/F*(^2P_{1/2})$ ratio in the fluorine atom beam source. But because of the resonance factor in REMPI, and complex data processing in Stern–Gerlach magnetic analysis, it is difficult for both of them to accurately obtain the $F(^2P_{3/2})/F*(^2P_{1/2})$ ratio in the fluorine atom beam source. In *Faraday Discussion* in 1999, the $F(^2P_{3/2})/F*(^2P_{1/2})$ ratio in the fluorine atom beam source was a focus of the general discussion on the non-adiabatic effects in F + H_2 reaction [18]. If this value is unknown, it is meaningless to compare the integral cross sections from experiments and theoretical calculation. As a matter of fact, Ruscic and coworkers carried out measurement of the autoionization spectra of the spin–orbit coupling ground and excited state fluorine atoms back to 1984 [19]. However, this approach has been ignored as a method to measure the $F(^2P_{3/2})/F*(^2P_{1/2})$ ratio in the fluorine atom beam source. In this thesis, we adopted this single-photon autoionization method to measure the $F(^2P_{3/2})/F*(^2P_{1/2})$ ratio in the fluorine atom beam source.

To measure the $F(^2P_{3/2})/F*(^2P_{1/2})$ ratio in the beam, we used single-photon absorption through autoionizing states of the F atom using vacuum ultraviolet

(VUV) light near 682 Å at the National Synchrotron Radiation Laboratory in Hefei, China. The F atoms are generated using a double-stage pulsed discharge method [20]. After ejecting through the nozzle and ∼117.2 mm flight (similar to the distance between beam source and cross region in our crossed molecular beam experiments), the F atoms perpendicularly intersect with the synchrotron light and absorbed ∼18.2 eV VUV photon. The F atoms were then excited from F(^{2}P$_{3/2}$) and F*(^{2}P$_{1/2}$) to two autoionization states: (2p^4(^{1}S$_0$)3s, ^{2}S$_{1/2}$) and (2p^4(^{1}D$_2$)4s, ^{2}D$_{3/2,5/2}$). After autoionization, the fluorine ions are detected by reflection TOF mass spectrometer. The spectra were taken by collecting the F ions when we scanned the wavelength of the VUV synchrotron light, which passes through the F/F* beam. During the whole experiment, the synchrotron light intensity was monitored by the photodiode and recorded by a digital picoammeter, to normalize the collected spectra.

Figure 4.3 shows the autoionization spectra of F(^{2}P$_{3/2}$) and F*(^{2}P$_{1/2}$) in the F atom beam source. The F atom beam was generated by expanding a mixture of 50 % NF$_3$/2.5 % F$_2$/47.5 % He (back pressure 4 atm) through a double-stage discharge. Among the four peaks in the spectra, two peaks correspond to fluorine atom spin–orbit ground state F(^{2}P$_{3/2}$), while the other two correspond to excited state F*(^{2}P$_{1/2}$). By measuring the areas under the peaks corresponding to excitation of the F and F* to the same autoionization state, the F and F* ratio in the beam can be determined, assuming F(^{2}P$_{3/2}$) and F*(^{2}P$_{1/2}$) have equal ionization efficiency. In the measurement, the ratio of F/F* was determined to be 4.8. Accordingly, the electron temperature was estimated to 663 K. The fluorine atom speed was estimated to approximately 1,080 m/s, based on the distance between discharge region and ionization region, and the time delay of discharge and ion extraction. During the adiabatic expansion, the electron temperature is substantially equal to the beam source temperature, according to the relationship between the beam source temperature and the molecular beam speed. That is to say, F* in the molecular beam was fully relaxed, and the ratio of F/F* has reached thermal equilibrium. In earlier literature, the F(^{2}P$_{3/2}$)/F*(^{2}P$_{1/2}$) ratio was assumed to be 2:1 (without relaxation or

Fig. 4.3 The autoionization spectrum of F(^{2}P$_{3/2}$) and F*(^{2}P$_{1/2}$) in the F atom beam source. The open circles are the experimental data, while the solid lines are the Gaussian fitted lineshapes for the four transitions. From [25], reprinted with permission from AAAS

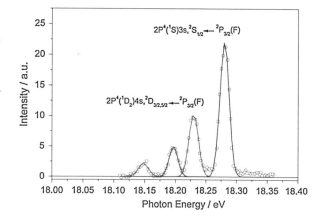

degeneracy ratio of electronic states) [6, 21]. The assumption may have a problem. Of course, the ratio varies when using different ways to generate F atoms. We also measured the F(^{2}P$_{3/2}$)/F*(^{2}P$_{1/2}$) ratio in fluorine atom beam generated by expanding pure NF$_3$ (back pressure 6 atm) and a mixture of 5 % F$_2$/95 % He (air supply pressure 5.3 atm), which were 4.7 and 10.7, respectively [22].

Single-photon VUV autoionization spectra allow us to get the F(^{2}P$_{3/2}$)/F*(^{2}P$_{1/2}$) ratio in fluorine atom beam, which is necessary to study the non-adiabatic effects in the reactions involving F(^{2}P$_{3/2}$) and F*(^{2}P$_{1/2}$), experimentally and quantitatively, and to compare with theoretical calculation.

4.3   Experimental Results and Analysis of Crossed Molecular Beam

4.3.1   Time-of-Flight Spectra of the D Atom Product

We had probed at many laboratory angles in the range from 0° to 180° varied in ~10° intervals in the center-of-mass frame at each collision energy. We found that F + D$_2$ reaction product signal is mainly sideways–backward. There is no forward signal. So, we detected at 10 laboratory angles ranging from sideways to backward at each collision energy. Not all of the TOF spectra are shown here, but only two of them are discussed representatively. Figure 4.4 shows two typical TOF spectra of the D atom product from the F(^{2}P$_{3/2}$)/F*(^{2}P$_{1/2}$) + D$_2$(j = 0) reaction at collision energies of (a) 0.55 kcal/mol at a laboratory angle of 110° and (b) 1.12 kcal/mol at a laboratory angle of 125°. The two laboratory angles correspond to the D atom product backward scattering at corresponding collision energies. From the TOF spectra, reaction products from the ground state F(^{2}P$_{3/2}$) and excited state F*(^{2}P$_{1/2}$) can be clearly identified. Fully separation is seen in the TOF spectra for reactions of F and F* that produce DF in $v' = 4$, which is the highest energetically allowed product vibrational level. There are partially overlap for reactions that produce DF in $v' = 3$ and $v' = 2$. Fortunately, the rotational population is relatively cold, so the main structure is clear. By fitting, we expect that the F* reaction should produce DF($v' = 2, 3$) with slightly hotter or the same rotational distributions, and the shape of the rotational population is relatively smooth, which is consistent with theoretical calculation, as shown in Fig. 4.5.

4.3.2   Differential Cross Section

At all collision energies, quantum-state resolved DCSs of the DF product in the CM frame ($\theta_{cm} = 0°$–180°) were determined by fitting the above experimental results. At 12 collision energies we studied, the DCSs of F(^{2}P$_{3/2}$) + D$_2$(j = 0) and F*(^{2}P$_{1/2}$) + D$_2$(j = 0) are similar. Reaction products are backward scattering,

Fig. 4.4 TOF spectra of the D atom product from the $F(^2P_{3/2})/F*(^2P_{1/2}) + D_2(j = 0)$ reaction at collision energies of (*upper panel*) 0.55 kcal/mol at a laboratory angle of 110° and (*lower panel*) 1.12 kcal/mol at the laboratory angle of 125°. From [25], reprinted with permission from AAAS

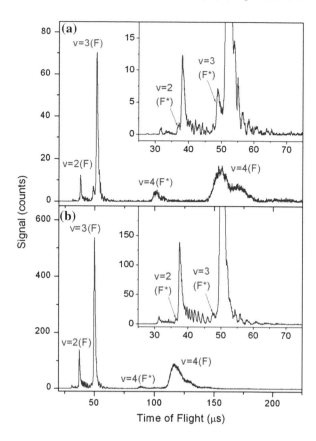

so here we discuss DCSs at 0.48 kcal/mol as an example. Figure 4.6 displays the topology of these DCSs at $Ec = 0.48$ kcal/mol and compares these with the results of theoretical calculations. In Fig. 4.6, (a) and (b) are experimental results, (c) and (d) are theoretical results from Alexander and coworkers, (a) and (c) are results from $F(^2P_{3/2}) + D_2(j = 0) \rightarrow DF + D$, and (b) and (d) are results from $F*(^2P_{1/2}) + D_2(j = 0) \rightarrow DF + D$. The main structure of this figure corresponds to DF product vibrational states $v' = 2, 3, 4$. Small peaks of each vibrational state correspond to rotational state of DF product. Most of the reaction products are backward scattering, which are direct reactions. Unlike other isotope reactions like $F(^2P_{3/2}) + H_2$ and $F(^2P_{3/2}) + HD$, the $F + D_2$ reaction does not have obvious reaction resonance phenomenon.

The theoretical simulations entailed time-independent, fully quantum, reactive scattering calculations on the LWA-78 PES by Alexander and coworkers. These calculations include explicitly all three PESs shown schematically in Fig. 4.2. LWA-78 PES overcomes the drawback on the ASW PES developed by Werner and coworkers [10, 23, 24], namely the underestimation by ~ 0.7 kcal/mol of the $F + H_2$ reaction exoergicity. Although this error is much smaller than the typical accuracy of ab initio calculations of PESs, it is a large fraction of the $F + H_2$ (D_2)

Fig. 4.5 The calculated relative rotational distributions for the DF($v' = 2$; *upper panel*) and DF($v' = 3$; *lower panel*) products at a collision energy of 0.7395 kcal/mol, in reaction of F (*filled circles*) and F* (*open circles*). The relative rotational distributions are defined as the integral cross sections (ICS) for production of a particular v', j' level divided by the sum of the ICS's for production of all j' levels in the given vibrational manifold. From [25], reprinted with permission from AAAS

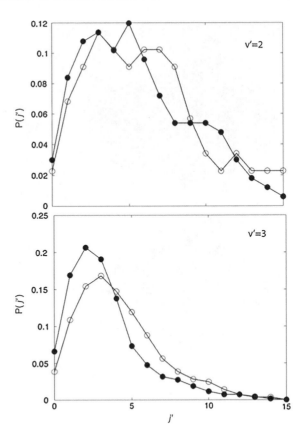

collision energy in the regime of low energies where one might expect to see the most noticeable evidence of non-adiabatic reaction dynamics. It is reasonable to believe that application of LWA-78 PES can get better results.

Except for some small forward scattering in the theoretical simulations, which lies below the detection limit in our experiments, the overall agreement in the DCSs between experiment and theory is excellent, in terms of both scattering angle and DF product state and for reactions of both F and F*.

The overall topology of the DCSs displayed in Fig. 4.6 is very similar for F and F* reactions. The F* reaction can occur only by means of the non-adiabatic transfer of incoming collision flux from the PES of the non-reactive excited state of A′ reflection symmetry (labeled 2A′ in Fig. 4.2) to the PES of the ground (reactive) 1A′ state. Flux transfer will occur most readily at geometries for which the F atom spin–orbit splitting is equal to the energetic splitting between the PESs of the two states of A′ reflection symmetry. This seam of geometries occurs well out in the F + D₂ reactant valley. After undergoing the non-adiabatic 2A′ → 1A′ transfer, the initially non-reactive flux will be subject to the same forces that affect the ground state reactants, in particular over the barrier and out into the HF + H product arrangement. Because the product distributions are a manifestation of

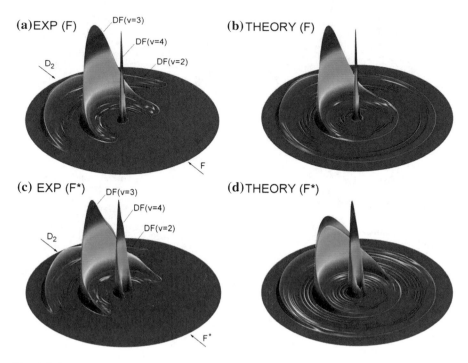

Fig. 4.6 Experimental (**a, b**) and theoretical (**c, d**) 3D D-atom product flux surface plots for the reaction F/F* + D$_2$ at 0.48 kcal/mol. From [25], reprinted with permission from AAAS

these forces, we would anticipate the DF + D DCSs arising from reaction of F and F* to be quite similar, as observed here.

The largest F* versus F difference can be seen in the DF ($v = 4$) products (Fig. 4.6). Because of the relative energetics (Fig. 4.2), for $Ec = 0.48$ kcal/mol, only the $j = 0$ and $j = 1$ DF rotational levels in $v = 4$ are energetically accessible in the F + D$_2$ reaction. The $j = 0$ products will emerge with a translational energy of only 0.004 kcal/mol. However, at the same collision energy, the first seven rotational levels in $v = 4$ are energetically accessible for the F* reaction, and the $j = 0$ products will emerge with a translational energy of 0.054 kcal/mol, substantially larger than that from the F reaction. It is for this reason that the DF ($v = 4$) DCSs from F and F* show the largest dissimilarity.

4.3.3 Reactivity Ratio Between F*($^2P_{1/2}$) and F($^2P_{3/2}$)

To assess the relative reactivity of the F and F* reactions, we determined the dependence on collision energy of the ICSs. These are the integrals of the corresponding DCSs over all scattering angles and so are a measure of the total reactivity at a particular Ec. To carry out this integration over scattering angle, we have chosen a set of laboratory angles that corresponds to the backward scattering direction for DF ($v' = 3$) product at a particular collision energy. We then measured TOF spectra at these laboratory angles and corresponding collision energies back and forward more than 10 times to put the ICS measurements at different collision energies on the same scale, such that the relative ICS could be determined experimentally by just integrating over all scattered products. At each collision energy, $F(^2P_{3/2}) + D_2(j = 0)$ and $F^*(^2P_{1/2}) + D_2(j = 0)$ were carried out under the same experimental conditions, so we multiply the ICS from $F^*(^2P_{1/2})$ reaction by $F(^2P_{3/2})/F^*(^2P_{1/2})$ ratio (=4.8), to get the $F^*(^2P_{1/2})/F(^2P_{3/2})$ reactivity ratio σ^*/σ at this particular collision energy.

Different collision energies were carried out under different experimental conditions, so they are not comparable. To obtain dependence on collision energy of relative integral cross sections, we studied dependence on collision energy of reaction product DF($v' = 3$) backward scattering signal. Figure 4.7 shows collision energy-dependent DCS for the backward scattering DF($v' = 3$) products summed over $j' = 0$–7. The signal increased with the increase in the collision energy.

Using collision energy dependence of backward scattering signal, collision energy dependence of $F(^2P_{3/2}) + D_2(j = 0)$ reaction relative integral cross section was obtained by adjusting integral cross section at each collision energy. At the same time, collision energy dependence of $F^*(^2P_{1/2}) + D_2(j = 0)$ reaction relative integral cross section and collision energy dependence of $F^*(^2P_{1/2})/F(^2P_{3/2})$ reactivity ratio σ^*/σ were also obtained.

In Fig. 4.8, the solid curves are the results of theoretical calculations from Alexander and coworkers, whereas the points indicate the experimental results. Figure 4.8a shows the experimentally determined ICSs for the F and F* reactions in the collision energy range between 0.25 and 1.2 kcal/mol. Over the entire set of collision energies, the experimental ICSs are scaled to the theoretical values by means of a single scaling factor for both the F and F* reactions. The reactive cross section for the B–O-allowed F atom reaction increases much more rapidly with

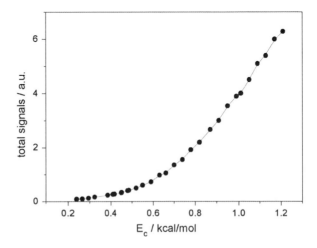

Fig. 4.7 Collision energy-dependent DCS for the backward scattering DF($v' = 3$) products summed over $j' = 0$–7

Fig. 4.8 a Collision energy dependence of the overall integral reactive cross sections, summed over product vibrational and rotational levels, for the F/F* + D$_2$($j = 0$) reactions for 0.25 ≤ E_c ≤ 1.2 kcal/mol; **b** the ratio of the cross sections shown in panel (**a**). The solid curves are the results of our theoretical calculations, whereas the points indicate the experimental results. The error bars indicate the range of measurement errors in the experiment. From [25], reprinted with permission from AAAS

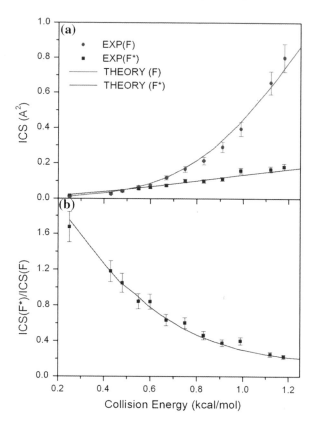

increasing Ec than that for the B–O forbidden reaction of F*, which was the qualitative prediction of theoretical calculations. The energy dependence of the ratio of the ICSs for the F and F* reactions is shown in Fig. 4.8b. The ratio σ^*/σ increases with decreasing Ec. At Ec lower than 0.5 kcal/mol, the F*(^{2}P$_{1/2}$) reactivity is greater than F(^{2}P$_{3/2}$), indicating F*(^{2}P$_{1/2}$) is predominant at low collision energies, and the non-adiabatic effects are obvious. The agreement with the predictions of the present scattering calculations based on the new PESs is remarkable. At the lowest Ec studied here, ∼0.25 kcal/mol, both experiment and theoretical calculations show that F* is ∼1.6 times more reactive than F. In this system at low collision energies, the non-adiabatic effects are important, and B–O approximation breaks down.

4.4 Conclusion

Through high-resolution, highly accurate crossed beam reactive scattering experiments, we studied the DCSs in F(^{2}P$_{3/2}$) + D$_2$($j = 0$) → DF + D and F*(^{2}P$_{1/2}$) + D$_2$($j = 0$) → DF + D reactions. We measured the F(^{2}P$_{3/2}$)/F*(^{2}P$_{1/2}$)

ratio in the fluorine atom beam using synchrotron radiation single-photon autoionization and obtained the $F(^2P_{3/2})/F^*(^2P_{1/2})$ reactivity ratio at different collision energies, as well as collision energy dependence of the $F(^2P_{3/2})/F^*(^2P_{1/2})$ reactivity ratio. We found that at high collision energies, the $F(^2P_{3/2})$ reactivity is greater than $F^*(^2P_{1/2})$, while at low collision energies, the $F^*(^2P_{1/2})$ reactivity is greater than $F(^2P_{3/2})$, and the B–O approximation breaks down. This is the first accurate experimental measurement of the non-adiabatic effects of this important system. The overall agreement between experiment and theory is excellent.

References

1. Tully JC (1974) Collisions of $F(^2P_{1/2})$ with H_2. J Chem Phys 60:3042–3050
2. Neumark DM, Wodtke AM, Robinson GN et al. (1985) Molecular beam studies of the $F + H_2$ reaction. J Chem Phys 82:3045–3066
3. Faubel M, Martinezhaya B, Rusin LY et al (1997) Experimental absolute cross-sections for the reaction $F + D_2$ at collision energies 90–240 meV. J Phys Chem A 101:6415–6428
4. Neumark DM, Wodtke AM, Robinson GN et al. (1985) Molecular beam studies of the $F + D_2$ and $F + HD$ reactions. J Chem Phys 82:3067–3077
5. Harper WW, Nizkorodov SA, Nesbitt DJ (2002) Reactive scattering of $F + HD \rightarrow$ HF(v,J) + D: HF(v,J) nascent product state distributions and evidence for quantum transition state resonances. J Chem Phys 116:5622–5632
6. Nizkorodov SA, Harper WW, Chapman WB et al. (1999) Energy-dependent cross sections and nonadiabatic reaction dynamics in $F(^2P_{3/2}, {}^2P_{1/2}) + n\text{-}H_2 \rightarrow$ HF(v,J) + H. J Chem Phys 111:8404–8416
7. Nizkorodov SA, Harper WW, Nesbitt DJ (1999) State-to-state reaction dynamics in crossed supersonic jets: threshold evidence for non-adiabatic channels in $F + H_2$. Faraday Discuss 113:107–117
8. Dong F, Lee SH, Liu K (2000) Reactive excitation functions for $F + p\text{-}H_2/n\text{-}H_2/D_2$ and the vibrational branching for $F + HD$. J Chem Phys 113:3633–3640
9. Alexander MH, Manolopoulos DE, Werner HJ (2000) An investigation of the $F + H_2$ reaction based on a full ab initio description of the open-shell character of the $F(^2P)$ atom. J Chem Phys 113:11084–11100
10. Alexander MH, Werner HJ, Manolopoulos DE (1998) Spin-orbit effects in the reaction of $F(^2P)$ with H_2. J Chem Phys 109:5710–5713
11. Zhang Y, Xie T-X, Han K-L et al. (2004) The investigation of spin-orbit effect for the $F(^2P) + HD$ reaction. J Chem Phys 120:6000–6004
12. Zhang Y, Xie T-X, Han K-L et al. (2003) Time-dependent quantum wave packet calculation for nonadiabatic $F(^2P_{3/2}, {}^2P_{1/2}) + H_2$ reaction. J Chem Phys 119:12921–12925
13. Zhang Y, Xie TM, Han KL (2003) Reactivity of the ground and excited spin-orbit states for the reaction of the $F(^2P_{3/2}), P^{21/2})$ with D_2. J Phys Chem A 107:10893–10896
14. Bischel WK, Jusinski LE (1985) Multiphoton ionization spectroscopy of atomic fluorine. Chem Phys Lett 120:337–341
15. Roth M, Maul C, Gericke KH et al (1999) State and energy characterisation of fluorine atoms in the A band photodissociation of F_2. Chem Phys Lett 305:319–326
16. Aquilanti V, Candori R, Cappelletti D et al. (1990) Scattering of magnetically analyzed F (^2P) atoms and their interactions with He, Ne, H_2 and CH_4. Chem Phys 145:293–305
17. Aquilanti V, Luzzatti E, Pirani F et al. (1988) Molecular beam studies of weak interactions for open-shell systems: the ground and lowest excited states of ArF, KrF, and XeF. J Chem Phys 89:6165–6175

18. Alexander M (1999) General discussion. Faraday Discuss 113:201–239
19. Ruscic B, Greene JP, Berkowitz J (1984) Photoionisation of atomic fluorine. J Phys B At Mol Physi 17:L79–L83
20. Ren ZF, Qiu MH, Che L et al (2006) A double-stage pulsed discharge fluorine atom beam source. Rev Sci Instrum 77:016102
21. Lee S-H, Dong F, Liu K (2004) A resonance-mediated non-adiabatic reaction: $F^*(^2P_{1/2}) + HD \rightarrow HF(v' = 3) + D$. Faraday Discuss 127:49–57
22. Che L (2008) thesis: studies on non-adiabatic effect and reaction resonance in the reaction $F(^2P) + H_2/HD/D_2$. In: Institute of Chemical Physics, CAS. Dalian China
23. Tzeng Y-R, Alexander MH (2004) Angular distributions for the $F + H_2 \rightarrow HF + H$ reaction: the role of the F spin-orbit excited state and comparison with molecular beam experiments. J Chem Phys 121:5812–5820
24. Tzeng YR, Alexander M (2004) Reactivity of the F spin-orbit excited state in the $F + HD$ reaction: product translational and rotational energy distributions. Phys Chem Chem Phys 6:5018–5025
25. Che L, Ren ZF, Wang XG et al (2007) Breakdown of the Born-Oppenheimer approximation in the $F + o\text{-}D_2 \rightarrow DF + D$ reaction. Science 317:1061–1064

Postscript

This thesis in Chinese was finished in July 2009, and the English version was completed in April 2013. In the last several years, the study on $F + H_2$ reaction system was further deepened in Yang's group in Dalian Institute of Chemical Physics, CAS. Xueming Yang, Donghui Zhang, and coworkers revealed that HF(v'=3) forward scattering in the $F + H_2$ reaction is not caused by either Feshbach or dynamical resonances, but caused predominantly by the shape resonance and slow-down mechanism [3]. They also measured the collision energy-dependent DCS of the reaction $F + HD \rightarrow HF + D$ with extremely high resolution, and observed the individual partial wave resonance with well-defined rotation in the transition state region [1]. In this thesis, we studied the dynamical effects of the H_2 rotation in the reaction resonance in the $F + H_2$ system. Recently, Yang's group realized as high as 91% population transferring from $v = 0$ to $v = 1$ of HD molecules using Stark-induced adiabatic Raman passage (SARP) scheme [2]. Their achievement provides new opportunities for future experimental studies on the dynamics of vibrational state molecules. Let us wait for the H_2-vibration effects in this benchmark system.

References

Dong W, Xiao C, Wang T et al (2010) Transition-State Spectroscopy of Partial Wave Resonances in the F + HD Reaction. Science 327:1501-1502

Wang T, Yang T, Xiao C et al (2013) Highly Efficient Pumping of Vibrationally Excited HD Molecules via Stark-Induced Adiabatic Raman Passage. The Journal of Physical Chemistry Letters 4:368-371

Wang XG, Dong WR, Qiu MH et al (2008) HF(v '=3) forward scattering in the F + H_2 reaction: Shape resonance and slow-down mechanism. Proceedings of the National Academy of Sciences of the United States of America 105:6227-6231

Printed by Publishers' Graphics LLC
DBT131112.15.19.102